April 29, 2003

This book is the property of

Hideo Stanley Yost

It was special-ordered

researched by Barnes & Noble Booksellers,

and was delivered to this author's

address in Wailuku, Maui, this

afternoon –

LETHAL YELLOWING: RESEARCH AND PRACTICAL ASPECTS

Developments in Plant Pathology

VOLUME 5

The titles published in this series are listed at the end of this volume.

Lethal Yellowing: Research and Practical Aspects

Edited by

C. OROPEZA
F. W. HOWARD

and

G. R. ASHBURNER

KLUWER ACADEMIC PUBLISHERS
DORDRECHT / BOSTON / LONDON

Library of Congress Cataloging-in-Publication Data

Lethal yellowing : research and practical aspects / edited by C. Oropeza, F.W. Howard, and G.R. Ashburner.
p. cm. -- (Developments in plant pathology ; v. 5)
Includes index.
ISBN 0-7923-3723-9 (HB : printed on acid free paper)
1. Palm lethal yellowing disease--Congresses. 2. Palm lethal yellowing mycoplasma-like organism--Control--Congresses.
I. Oropeza, C. II. Howard, F. W. III. Ashburner, G. R.
IV. Series.
SB608.P22L45 1995
634'.61932--dc20 95-37539

ISBN 0-7923-3723-9

Published by Kluwer Academic Publishers,
P.O. Box 17, 3300 AA Dordrecht, The Netherlands.

Kluwer Academic Publishers incorporates
the publishing programmes of
D. Reidel, Martinus Nijhoff, Dr W. Junk and MTP Press.

Sold and distributed in the U.S.A. and Canada
by Kluwer Academic Publishers,
101 Philip Drive, Norwell, MA 02061, U.S.A.

In all other countries, sold and distributed
by Kluwer Academic Publishers Group,
P.O. Box 322, 3300 AH Dordrecht, The Netherlands.

Printed on acid-free paper

Printed in the Netherlands

Contents

7. Future directions for the coconut industry

Foreword

When I received an invitation to attend the International Symposium on Lethal Yellowing being organised by the Centro de Investigación Científica de Yucatán (CICY), I was excited and a little nostalgic. During the 1970s, a series of similar symposia had been held under the auspices of the loosely-constituted "International Council on Lethal Yellowing" (ICLY). These were the years when the MLO cause for LY was first proposed, a vector was found, the disease was racing across mainland Florida, USA and it was suspected of having jumped to Cozumel. Analogous diseases were also reported to be spreading in Africa and elsewhere. The ICLY meetings, held approximately every two years, proved to be an immensely valuable forum for all involved in the research and control of LY. They attracted a very wide cross-section of scientists and practitioners working on LY, on related diseases, and on palms in general. Many participants of those ICLY meetings also attended this CICY Symposium. Unfortunately, during the 1980s, as countries learned to live with LY, most of the national and international funding for LY research dried up, and so did ICLY.

The present symposium is the only international meeting to have been devoted to LY since the last meeting of ICLY in 1979. Its convening in Mérida is timely. During the past few years, advances in biotechnology have provided diagnostic tools for the detection and quantification of LY that were undreamed of only a decade ago. México produces more copra than all other countries in the Americas combined and, with the threatened movement of LY out of the Yucatán Peninsula, that entire industry is at risk. Productive and landscape palms throughout Central America are also vulnerable. Staff at CICY, and the sponsors of this symposium are thus to be congratulated on renewing an international focus on LY.

The new developments presented at this Symposium are likely to lead to a resurgence of interest in LY and in similar diseases affecting palms. The disease is now known to affect over 30 palm species and the first non-palm, *Pandanus*, has been shown to be attacked by the LY agent. Of major importance to Africa, is the work reported here on the relationships between LY and diseases there which have similar symptoms. The ability to detect minute quantities of LY-MLOs in plant or insect tissues promises new insights into vector relationships, disease control through use of tolerant varieties and the possibility of seed or pollen transmission. These are but a few of the topics discussed at the Symposium.

There is a serious potential for the spread of vector-transmitted palm diseases, of which LY is but one, from their present fairly narrow geographical ranges to a global tropical distribution. For the developing world, the loss of coconuts alone would have an enormous impact, domestically as well as internationally; and there are many other economic and amenity palms which can and might be affected. It is hoped that this Symposium hosted by CICY is the first of a new set of international meetings on such palm diseases.

Peter Hunt
NRI/ODA Plant Protection Project, Belize

Introduction

The coconut palm is a very important plant in the tropical regions, both as cash crop and in subsistence agriculture. Although most coconut production is dedicated to copra and its subsequent oil extraction, the number of products which can be obtained from the coconut palm is seemingly limitless. However, the coconut industry has several problems which affect its productivity, particularly: the use of unimproved planting material, the old age of existing plantations and various pests and diseases. This book deals with the most severe of the diseases, lethal yellowing (LY). Lethal yellowing has killed millions of coconut palms in Latin America and the Caribbean alone, and which, together with related diseases in Africa and possibly India and Southeast Asia, poses a world-wide threat to coconut production. The papers in this volume were presented in a symposium entitled "Lethal Yellowing Research and Practical Aspects", which was held at Centro de Investigación Científica de Yucatán (Mérida, Yucatán, México) in November, 1993.

The location of the meeting in Mérida was particularly relevant since the region has been suffering from the effects of LY since the late 1970s. The disease has devastated the coconut groves in the Yucatán Peninsula and is now threatening the rest of the country and Central America. As such, there is much interest in the biology and control of the disease. Research on LY and coconut improvement has been performed by different institutions in México, in particular Centro de Investigación Científica de Yucatán (CICY) and Instituto Nacional de Investigaciones Forestales, Agricolas y Pecuarias (INIFAP), and efforts to fight LY are have been made by the Mexican phytosanitary authority, Dirección Nacional de Sanidad Vegetal (SARH) within a national-wide program.

The scope of the papers presented in this volume is wide so as to accommodate coconut improvement in general, coconut product diversification and a model of industry organisation, rather than just subjects directly related to LY. The rationale behind this breadth of subjects was to discuss the problem of lethal yellowing within the larger context of the coconut industry and to seek possible solutions from as wide a source as possible.

This volume includes introductory papers (Section 1) dealing with LY and a brief history of LY research which provides a general background for subsequent papers. Section 2 deals with various aspects of research which reveals how the pathogen kills palms, how the disease is spread and possibly why resistant palms are not affected. Section 3 includes papers on development of new methodology for the detection and diagnosis of lethal yellowing and its applications to basic research and practical usage.

Currently, the most efficient way to deal with LY is by planting resistant palms. This subject is dealt with in Section 4, as part of integrated management of the disease. Production of resistant hybrids and other aspects related to coconut genetic resources are presented in Section 5. This section also includes papers dealing with the reproductive biology and heredity of coconut palms, genetic markers and the safe movement of germplasm. They represent the current status of coconut resources and point to new directions for improvement of the crop.

The long generation time of coconut palms means that the production of improved planting materials cannot keep pace with the growing demand resulting from disease devastation or old age of plantations. Alternative techniques such as *in vitro* clonal propagation offer potential to speed up the process. Section 6 deals with new advances in that field.

Finally, the papers in Section 7 present examples of how the coconut industry can be restructured to cope with calamities such as LY and changing market profiles.

During the meeting these subjects were discussed and conclusions reached with the participation of experts representing the different fields of research, marketing and production. Although the Symposium was organised primarily with the objective of providing direction to combat the disease in México, the following conclusions are equally as relevant to other lethal yellowing-affected countries.

(a) The Symposium considered that there is a need to establish a program of field-testing the coconut germplasm available in México, and to import germplasm from overseas to test its agronomic performance. These tests should be done with numerous replicates to evaluate the germplasm's response to different environmental conditions. In the case of local tests in the LY-affected zone, tests should also be carried out with a view to evaluating disease resistance levels. The evaluation program should be performed in co-ordination with a genetic improvement program. Furthermore, the research experience that has been accumulated in other places, particularly Jamaica, should be taken into consideration.

(b) The Symposium considered that, with respect to the diagnostic methods for LY, the traditional methods should continue to be used, but molecular methods should be used to confirm the identity of the particular MLO.

Laboratories currently using the molecular diagnostic techniques should interchange material to corroborate results. Institutions not using molecular techniques should have the opportunity of staff training in laboratories where they are currently being practised.

(c) The Symposium considered that there is a need to identify the modes of dispersion of LY in México and to develop appropriate control measures. Additionally, it was considered very important to develop methods of controlled transmission of LY for challenging plants (in association with indexing) in the laboratory, nursery and in the field.

(d) The Symposium recognised the need for continuing the investigation of the *in vitro* culture of coconut palms to develop efficient protocols for micropropagation and embryo culture. With respect to this goal, the urgent need for major co-operation between institutions was highlighted.

(e) The Symposium recognised the need to reinforce consultative and extension activities within the coconut industry, with the aim of increasing production and productivity, particularly of the small producers.

(f) With the fall in the world copra and coconut oil prices, the Symposium was of the opinion that there is a need to promote the diversification of coconut products to ensure the continuing viability of the coconut industry. It was considered that to achieve this goal, it is necessary to complete an economic and social evaluation with respect to availability, potential technology and markets.

(g) Finally, the Symposium considered that, to achieve advances in all the aforementioned areas, it is fundamental to promote co-operation between Mexican and overseas institutions from which joint research projects can stem.

It is hoped that both the Symposium, and the publication of this volume, will maintain the current momentum in lethal yellowing research, and perhaps stimulate further research on coconut palms, an orphan crop with a thousand uses.

Carlos Oropeza
Bill Howard
Roger Ashburner

Acknowledgements

Both the Symposium on lethal yellowing and the production of this book would not have been possible without support from the following organisations: the Government of México through the Consejo Nacional de Ciencia y Tecnología and the Dirección General de Sanidad Vegetal of the Secretariat of Agriculture and Water Resources, the Government of the state of Yucatán through its Secretariats of Ecology and Tourism, The British Council of México, Instituto Interamericano de Cooperación para la Agricultura, Hacienda Oxtapacab and Centro de Investigación Científica de Yucatán. We also acknowledge the assistance of many individuals, too numerous to mention.

Lethal yellowing

JOSE ARELLANO and CARLOS OROPEZA
Centro de Investigación Científica de Yucatán, Apartado Postal 87, Cordemex, 97310 Mérida, Yucatán, México

1. Introduction

Lethal Yellowing (LY) is a devastating disease that affects more than thirty palm species including coconut [49]. LY has been reported throughout the Caribbean Region [29, 49] where its effects have been more conspicuous on coconut palms than other species because of their abundance and integral part of the landscape. In the last fifty years LY epidemics in the Caribbean Region have killed millions of palms and have resulted in serious economic losses to hundreds of thousands of families whose incomes depend directly or indirectly on coconut. Extensive research efforts have endeavoured to find ways to control LY over the last three decades. Unfortunately, effective methods of LY control have not been forthcoming. The purpose of this paper is to provide general information on LY, and most of the subjects dealt with here are discussed more extensively in subsequent papers. Where required, the reader will be referred to previously published reviews.

2. Distribution

Reports of dying coconut palms exhibiting LY-like symptoms date from the last century in the Cayman Islands [40, 42, 44, 51, 52], Cuba [14, 40] and Jamaica [22, 44]. During the 20th century, extensive damage by LY was reported in Cuba during the 1940s [3] and in Jamaica during the 1960s and 1970s [3, 37], and LY was also reported in the Dominican Republic [10, 57], Haiti [41], Bahamas [41] and in Florida in the USA [43]. An association of mycoplasmalike organisms (MLO) with LY was reported in Jamaica [1] and Florida, USA [65, 66], and with a similar or identical disease of date palms (*Phoenix* spp) in Texas, USA [48]. More recent outbreaks of the disease have been reported in Mexico's Yucatán Peninsula

C. Oropeza et al. (eds.), Lethal Yellowing: Research and Practical Aspects, 1–15.

[5, 49] and in Belize [21]. Reports on the possible presence of LY in other countries in the region have appeared but have yet to be confirmed [49]. Occurrence of MLO-associated lethal yellowing diseases (LYD) similar to LY have been reported from the West African countries of Ghana [13], Togo [13], Cameroon [16] and Nigeria [4, 20] as well as Tanzania in East Africa [61]. LYD from West Africa shows similarities in symptoms [62] and epidemiology [47] to LY in the Caribbean. In the case of LYD from Tanzania, there are also similarities in symptoms but there are differences in epidemiology [60]. There is also evidence to indicate that the causal agent in Tanzania is similar to that that causes LY but is not identical [25] (see below).

3. LY spread

3.1 Patterns of spread

Two types of spread of LY have been reported in Jamaica [7, 9] and Florida [49]. One type involves a local centre of infection that appears in one or two palms. This is followed by new cases appearing at random around the initial centre, extending local spread. The second type is a jump spread followed by local spread. The distance of these jumps varies from a few to 70 kilometres [7]. Similarly, these two patterns of spread of LY have been observed in México (Escamilla pers. comm.).

McCoy [47] noted that the rate of long distance spread in Jamaica appeared to be slower than in Florida. It took more than 60 years to cover the distance between the west and east end of the island, whereas it jumped from Miami to Palm Beach and Naples in Florida and to Nassau in the Bahamas within 3 years. McCoy [47] considered that the mountainous terrain of Jamaica probably contributed to the slower rate of long distance dispersal. Florida has no barriers to air-borne dispersal. In México, LY was first observed in the Cozumel-Cancun area in 1979 and has spread through the Yucatán Peninsula to the Campeche-Tabasco border about 900 km away in 15 years (Escamilla et al., this volume). The Yucatán Peninsula also has no barriers to airborne dispersal. Furthermore, it is currently approaching the Isthmus of Tehuantepec which is a geographic corridor between the Gulf of México and the Pacific Ocean, also with no barriers to air-borne dispersal. So, the threat of LY reaching the Pacific coast of México is growing alarmingly. This is the most important coconut growing region in México.

3.2 Rate of spread

The first cases of LY on the Florida mainland were found in Dade County in late 1971 [63]. The estimated coconut palm population was 350 000 before the appearance of LY and accurate records of the losses by LY were kept. At the beginning of the epidemic, 50 diseased palms were evident and, by the end of 1972, 2 000 had been affected. By autumn 1973, this figure had increased to nearly 20 000. Loses in Dade County approached 50% of the original population by the end of 1974 and 75% by the end of 1975 [47].

The LY tree-to-tree spread in Dade County was calculated by McCoy *et al.* [49]. He reported that, in 1971, in the first 8 months when only a small portion of the area was affected, each infected palm served to inoculate 4.6 new palms. In 1973, when the logarithmic stage of spread was well underway, each infected palm served to infect 9.3 new palms. McCoy *et al.* [49] considered this increase as a result of the greater predominance of inoculum in relation to the remaining uninfected palms. He reported that, in general, all the coconut palms were dead or dying approximately 2 years after the first cases of LY were evident, in any area of a few city blocks.

The type of locality was also found to affect the rate of the spread (expressed as r values) of LY [47]. In Florida, the highest rate (0.42 per unit per month - pupm) was found in inland groups of palms receiving regular irrigation and fertilisation and the lowest rate (0.15 pupm) occurred adjacent to the ocean, even with high maintenance. Intermediate r values (0.21 pupm) were calculated in inland sites receiving minimal maintenance. This was found to be similar in Jamaica (0.30 pupm). In addition, it is interesting to note that in breeding plots in Jamaica, r calculated from resistance data for Jamaica Tall (JT) was 0.06 pupm. The JT palms were in mixed planting with resistant ecotypes. McCoy *et al.* [49] stressed that the rapid spread of LY represented a threat to coconut and date palm production world-wide and that this is emphasized by the occurrence of similar diseases in Africa.

Figure 1. Progressive development of lethal yellowing symptoms: premature drop of most of the fruit (A); blackening of unemerged (B) and emerged (C) inflorescences; yellowing of lower (D), middle (E) and upper (F) leaves; and bare trunk or "telephone pole" (G).

4. Symptoms

4.1 Symptoms in coconut palms

The first visual symptom of the disease in bearing coconut palms affected by lethal yellowing is the premature drop of most of the fruit regardless of their developmental stage (Fig. 1A). The next symptom to appear is the blackening of new inflorescences (Fig. 1B, 1C). This symptom shows as the inflorescence emerges from the spathe. The first affected inflorescences usually show partial necrosis, but, as the disease progresses, newer inflorescences show more extensive necrosis. Most of the male flowers are dead and no fruit are set on those affected inflorescences.

Yellowing of the leaves usually starts after necrosis has developed in more than two inflorescences. The pattern of leaf discoloration due to LY is more rapid than that for normal leaf senescence. The first leaves to turn yellow are the oldest (lower) ones (Fig. 1D), then yellowing advances upwards; affecting the younger middle (Fig. 1E) and finally the upper leaves (Fig. 1F). Yellow leaves turn brown, desiccate and die. They remain hanging for a few days before falling. Eventually, the whole crown perishes, leaving a bare trunk or "telephone pole" (Fig. 1G). The spear leaf is also affected, developing a rotten area and it occasionally dies when other upper leaves are still green (Fig. 1H). McCoy *et al.* [49] reported that, in Florida, palms with LY occasionally showed a single yellow leaf in midcrown. Based on this progressive pattern of symptom development, rating systems were devised by Eden-Green [17] and McCoy *et al.* [49], the latter and the most commonly used rating system is shown in Table 1.

In addition to these symptoms in above ground parts, roots of coconut palms also show necrosis which becomes more extensive as the disease progresses [18]. The growth of palms is also affected by LY. Dabek [12] reported that 80 days prior to symptom appearance, growth was stimulated up to almost 40% above control, then there was a gradual decline over 50 days before growth failed completely. When LY affected coconut palms are studied using field equipment and laboratory techniques, physiological and biochemical symptoms can be detected, such as the reduction in stomatal conductance. This subject is dealt with in more detail in Oropeza *et al.* (this volume).

Table 1. Rating of lethal yellowing development in coconut [49]

Category	Stage	Symptoms
Symptomless	0	Healthy or incubating
Primary[1]	1	Nutfall only
	2	One necrotic inflorescence
	3	Two or more necrotic inflorescences
Yellowing	4	Yellowing in lower leaves only
	5	Yellowing in lower and middle leaves
	6	All leaves yellowed, spear leaf good
Dying	7	Spear leaf dead, some green leaves left
	8	Spear leaf dead, all leaves yellowed
	9	Palm dead (telephone pole)

(1) May or may not have a yellow flag leaf in centre of crown

The time that elapses from probable time of acquisition of the pathogen to appearance of first symptom (nutfall) have been variously reported as follows: in mature bearing palms, 230-450 days [56] and 210-450 days [26]; for young non-bearing palms, at least 240-270 days [11]. The time between probably initial infection and death of mature palms has been reported as 3-6 months [23] and 4-5 months [46]

4.2 Symptoms in other palm species

Symptoms of LY in other palms have been described by McCoy *et al.* [49]. They observed that early symptoms, premature dropping and necrosis of new inflorescences, are generally similar to those observed in coconut palms. However, differences may occur in the stage at which spear leaf necrosis appears and the form of foliar discoloration. Two patterns of leaf discoloration have been described [49, 64]. Leaves turn a golden-yellow before dying in species such as coconut palm, gebang palm (*Corypha elata*), Fiji fan palm (*Pritchardia pacifica*), arikury palm (*Syagrus schizophylla*), windmill palm (*Trachycarpus fortunei*), princess palm (*Dictyosperma album*) and spindle palm (*Hyophorbe verschaffeltii*). In other susceptible palms, leaves turn brown rather than yellow. These include, among others: Manila palm (*Veitchia merrillii*), palmyra palm (*Borassus flabellifera*), cabada palm (*Chrysalidocarpus cabadae*) and date palms (*Phoenix* spp.). Whether in yellowing or non-yellowing palms, discoloration generally advances from the oldest to the youngest leaves in the crown [49].

5. Causal agent

Initial studies unsuccessfully investigated bacteria, fungi, viruses, nematodes and abiotic factors as causing LY [49]. MLO were first found to be associated with some plant diseases during the 1960s [15, 38]. These results then sparked the search for MLO in LY-affected palms. In 1972, three groups independently reported the occurrence of MLO in the phloem of coconut palms showing LY symptoms [1, 54, 26] and this result has been further confirmed [53, 65, 66]. A cause-effect relationship between MLO and LY was supported by the differential response of LY diseased palms to antibiotics. LY palms treated with penicillin showed no beneficial response whereas symptom remission occurred when they were treated with oxytetracycline [37, 45]. Evidence has been reported that supports an MLO aetiology for LYD in West Africa [16, 20] and East Africa [62]. A recent study using a combination of DNA probes and polymerase chain reaction (PCR) compared MLOs associated with Caribbean LY and African LYD [25]. This study demonstrated that the pathogens associated with Caribbean LY and coconut LYD in both West and East Africa are generally related.

6. Transmission

According to the patterns of spread of LY, it was first hypothesised that it was transmitted by flying insects or birds [40]. Researchers in Cuba [3] and Jamaica [9, 51] suggested that the disease was probably caused by a virus and transmitted by an insect species of the order Homoptera. On the assumption that LY was caused by a virus, several groups of insects were tested as vectors, such as leafhoppers, whiteflies, aphids, scales, mealybugs and thrips. A review of these experiments was published by Johnson and Eden-Green in 1978 [39]. When the causal agent of LY was discovered to be an MLO [1, 26, 37, 45, 53, 54], the search concentrated on species of Auchenorrhyncha, the sub-order of the Homoptera to which most vectors of MLO-associated diseases belong [70]. Surveys conducted in LY-affected areas in Jamaica yielded five species of fulgorids [58, 59] and in Florida two fulgorids and one membracid [27, 31, 32]. The only common species found in coconut palms in both locations was *M. crudus*. In addition, the apparent rate of spread of LY decreased in areas where *M. crudus* populations were reduced by insecticide treatment [31] and populations of *M. crudus* were 40 times higher in heavily affected areas than in LY-free areas in Florida [28]. As a result of these reports, *M. crudus* has been extensively tested as a vector for LY (Howard, this volume).

Initial LY transmission tests were conducted in Jamaica [19] and in Florida [69], but transmission was not achieved. In a second series of tests in Florida, Howard *et al.* [33, 34] used larger numbers of insects and longer incubation times than in previous experiments to favour transmission. They were successful in transmitting LY to caged *C. nucifera*, *Veitchia merrillii* and *Pritchardia thurstonii*. Every month, approximately 850 wild *M. crudus* insects were fed into each cage over a 34 month period. Palms in cages where insects were not introduced remained healthy.

Taken together, the epidemiology data and the transmission experiments strongly support *M. crudus* as the vector of LY in the Caribbean Region. However, the presence of the LY-MLO within *M. crudus* has yet to be demonstrated. This was attempted by Waters [72] using electron microscopy but no evidence was forthcoming. Molecular techniques such as DNA probes and PCR are potentially useful for this purpose (Harrison *et al.*, this volume).

7. Host range

LY is commonly associated with coconut palms but it is known to affect at least 30 additional palm species [49, 66, 68]. With one exception, there is no evidence that plants other than palms serve as hosts of the LY-MLO. The exception is *Pandanus utilis* which is affected by a lethal MLO-associated disease in Florida [67]. This MLO was confirmed as the LY MLO by Harrison *et al.* (this volume) using DNA probes.

It appears that young palms (up to three years old) are not susceptible or are rarely affected by LY under field conditions [6, 7, 51]. However, under experimental conditions, young palms were found to develop LY. Howard *et al.* [36] reported that LY developed in young (less than 2 years old) palms of *C. nucifera*, *Pritchardia pacifica* and *Trachycarpus fortunei* when kept in cages with *Myndus crudus* present. A *T. fortunei* palm in this study was only 14 months old when it developed LY symptoms.

The possibility of symptomless hosts of LY has intrigued researchers. Howard *et al.* [35] searched for species that might possibly served as symptomless hosts of LY. They looked for common isolated LY-affected localities. Based on these results, they suggested three graminaceous species that are hosts of *M. crudus*, *Stenotaphrum secundatum*, *Cynodon dactylon* and *Cenchrus* spp.. The occurrence of the LY-MLO in these plants could be tested with the current molecular methods developed by Harrison *et al.* (this volume).

8. Diagnosis and detection

The visual symptom progression of LY allows for a tentative diagnosis. However, confirmation of the disease requires the detection of the LY MLO. Occurrence of MLO in diseased tissue can be achieved directly by observation of the MLO cells in preparations of diseased tissue using transmission electron microscopy or indirectly by observation of the DNA of the MLOs, using the DAPI fluorochrome, with an epifluorescence microscope. These techniques are not specific since they cannot differentiate one MLO from another. However, diagnosis can be confirmed by the differential responses of diseased plants to the antibiotics penicillin and oxytetracyline, which provide evidence of cause-effect relationship of MLO occurrence and LY.

Recent progress in molecular techniques has been made and methodology is currently available for the detection of LY using DNA probes and PCR. Harrison *et al.* [24] have developed LY-specific DNA probes and primers for PCR. See Harrison *et al.* (this volume) for a more comprehensive review of diagnosis and detection.

9. Control methods

Despite decades of research, a cure for LY is not yet available, but measures may be taken to attempt to reduce its rate of spread. Current and potential methods include quarantine, chemotherapy, vector control, sanitation and the use of resistant varieties. The latter appears to be the most effective management tool. A more in depth review on integrated pest management is presented by Been (this volume).

9.1 Vector control

The nymphs of *M. crudus* live and feed in the thatch layer and on the roots of certain grasses [30,71,72]. Accordingly, control of the grass hosts of this insect is a potential method of controlling it (see Howard, this volume). An alternative method of controlling the vector is by using insecticides. In tests carried out by Howard and McCoy [31], and Reinert [55], *M. crudus* populations were reduced by insecticide treatment, but not sufficiently to be recommended for practical purposes [49].

9.2 Chemotherapy

When the MLO nature of the causal agent of LY was discovered, antibiotics were immediately tested in Florida [45] and Jamaica [37]. It was found that tetracycline group antibiotics suppressed symptom development if applied before expression of systemic foliar yellowing. Chemotherapy has been successfully used for treating ornamental palms used for ornamental purposes, but chemotherapy was not feasible for commercial plantations because of its high cost and perceived health risks.

9.3 Sanitation

According to McCoy *et al.* [49], eradication of diseased palms could be useful in slowing the spread of LY if practised in the early stages of the outbreak. He noted that a major drawback of this practice is that LY has a long latent period.

9.4 Quarantine

LY spreads rapidly in any local area where it has become established. The disease also jumps large distances to establish new infection sites. However, the greatest majority of new cases occur within 100 metres of an established case of disease

[50]. Under these conditions, the quarantine of infested areas would retard disease spread.

9.5 Use of resistant varieties

Replanting with resistant palms has proved to be the most efficient way to deal with LY. Trials evaluating resistance to LY have been performed in Jamaica [2]. Resistance was found in some ecotypes, including the 'Malayan Dwarf' (MD) varieties, of which the yellow variety was subsequently used as a parent in F_1 hybrid production. The hybrids produced with MD as one parent had a level of resistance sufficiently high to be used for commercial planting. Based on these findings, replanting in Jamaica using MD and the 'Maypan' hybrids (MD x 'Panama Tall') has proven successful (Been, this volume). From a commercial perspective, 'Maypan' offers several advantages: it is resistant to LY, precocious and highly productive [2]. Therefore, its use has been extensively recommended. For ornamental purposes, several resistant species have been recommended [49].

Acknowledgements

The personal communication of José Escamilla of CICY, México is gratefully acknowledged.

References

1. Beakbane AB, Slater CHW and Posnette AF (1972) Mycoplasmas in the phloem of coconut, *Cocos nucifera* L., with lethal yellowing disease. J Hort Sci 47:265
2. Been BO (1991) Observations on field resistance to lethal yellowing in coconut varieties and hybrids in Jamaica. Oléagineux 36:9-11
3. Bruner SC and Boucle L (1943) La enfermedad conocida como "enfermedad del cogollo del cocotero en Cuba". Rev Agric 26:132-141
4. Bull RA (1955) Bronze leaf wilt of coconut palms in Nigeria. J West Afric Inst Palm Res 3:70-72
5. Cardeña R, Villanueva MA, Santamaría JM and Oropeza C (1991) Presence in Yucatán of mycoplasmalike organisms in *Cocos nucifera* L. palms showing lethal yellowing disease symptoms. Can J Plant Path 13:135-138
6. Carter W (1964) Present status of research of lethal yellowing disease of coconut palm in Jamaica. FAO Plant Protect Bull 12:67-69
7. Carter W (1966) Susceptibility of coconut palm to lethal yellowing disease. Nature 212:320
8. Carter W, Latta RK and Suah JRR (1965) The symptoms of lethal yellowing disease of coconut palms. FAO Plant Protect Bull 13:49-55

9. Carter W and Suah JRR (1964) Studies on the spread of lethal yellowing disease of the coconut palm. FAO Plant Protect Bull 12:73-78
10. Ciferri R and Cicarone A (1949) Observaciones sobre la enfermedad de la hoja broceada del cocotero en Venezuela. Rev Fac Agron Medellin 9:17-26
11. Dabek AJ (1974) Biochemistry of coconut palms affected lethal yellowing, a disease associated with mycoplasma-like organisms. Phytopath Z 81:346-353
12. Dabek AJ (1975) The incubation period, rate of transmission and effect on growth of coconut lethal yellowing disease in Jamaica. Phytopath Z 84:1-9
13. Dabek AJ, Johnson CG and Harries HC (1976) Micoplasma-like organisms associated with Kaincope and Cape St. Paul wilt diseases of coconut palms in West Africa. PANS 22:354-358
14. De La Torre C (1906) La enfermedad de los cocoteros. Rev Fac Let Cien 2:269-281
15. Doi Y, Teranaka M, Yora K and Asuyama H (1967) Mycoplasma or PLT group-like microorganisms found in the phloem elements of plants infected with mulberry dwarf, potato witches' broom, aster yellows or Paulownia witches' broom. Ann Phytopath Soc Japan 33:259-266
16. Dollet M, Giannotti J, Renard JL and Ghosh SK (1977) Étude d'un jaunissement létal des cocotiers au Cameroun: la maladie de Kribi. Observations d'organismes de type mycoplasmes. Oléagineux 32:317-322
17. Eden-Green SJ (1976) Disease scoring scale for lethal yellowing. In: 16th Rep Res Dept, pp. 39-41, Coconut Industry Board, Kingston
18. Eden-Green SJ (1979) Root symptoms in coconut palms affected by lethal yellowing disease in Jamaica. FAO Plant Protect Bull 4:119-122
19. Eden-Green SJ (1979) Attempts to transmit lethal yellowing disease of coconuts in Jamaica by leafhoppers (Homoptera: Cacadelloidea). Trop Agric 56:185-192
20. Epko EN and Ojomo EE (1992) The spread of lethal coconut diseases in West Africa: incidence of Awka disease (or bronze leaf wilt) in the Ishan area of Bendel state in Nigeria. Principes 34:143-146
21. Escamilla JA, Oropeza C, Harrison N, Alpizar L, Cordova I and Islas-Flores I (1994) Evaluación de sondas moleculares de ADN para el estudio de organismos tipo micoplasma causantes del amarillamiento letal. Reporte de Proyecto, CONACyT, México
22. Fawcet W (1891) Report on the coconut disease at Montego Bay. Bull 23 Bot Dept, Jamaica
23. Grylls NE and Hunt P (1971) A review of the study of the aetiology of coconut lethal yellowing disease. Oléagineux 26:311-315
24. Harrison NA, Burner R, Cox R, Tsai JH and Richardson P (1992) DNA probe for detection of mycoplasmalike-organisms associated with lethal yellowing of palms of Florida. Phytopath 82:216-224
25. Harrison NA, Richardson PA, Jones P, Tymon AM, Eden-Green SJ and Mpunami AA (1994) Comparative investigation of MLOs associated with Caribbean and African coconut lethal decline diseases by DNA hybridization and PCR assays. Plant Dis 78:507-511
26. Heinze KG, Schuiling M, Romney DH (1972) The possible cause of lethal yellowing disease of coconut. FAO Plant Protect Bull 20:58-68
27. Howard FW (1980) Attractiveness of date and coconut palms to *Myndus crudus* and other homopterans. Proc Florida State Hort Soc 93:199-201
28. Howard FW (1980) Population densities of *Myndus crudus* Van Duzee (Homoptera: Cixciidae) in relation to coconut lethal yellowing distribution in Florida. Principes 24:174-178

29. Howard FW (1983) World distribution and possible geographic origin of palm lethal yellowing disease and its vectors. FAO Plant Protect Bull 31:101-112
30. Howard FW (1990) Evaluation of grasses for cultural control of *Myndus crudus*, a vector of lethal yellowing palms. Entomol Exp Appl 56:131-137
31. Howard FW and McCoy RE (1980) Reduction in spread in mycloplasma-like organism associated lethal decline of the palm *Veitchia merrillii* by the use of insecticides. J Econ Entomol 73:268-270
32. Howard FW and Mead FM (1980) A survey of Auchenorrhyncha (Insecta: Homoptera) associated with palms in southern Florida. Trop Agric 57:145-153
33. Howard FW, Norris RC and Thomas DL (1982) Evidence of transmission of palm lethal yellowing agent by a planthopper, *Myndus crudus*. Trop Agric 60:168-171
34. Howard FW and Thomas DL (1980) Transmission of palm lethal decline to *Veitchia merrillii* by planthopper, *Myndus crudus*. J Econ Entomol 73:715-717
35. Howard FW, Tiltmant JR, Burch DG and Hall D (1984) A field study to determine potential host plants of lethal yellowing agent. Rep FL-84-1, Fort Lauderdale Res Educ Cent, University of Florida
36. Howard FW, Williams DS and Norris RC (1984) Insect transmission of lethal yellowing to young palms. Int J Entomol 26:331-338
37. Hunt P, Dabek AJ and Schuiling M (1974) Remission of symptoms following tetracycline treatment of lethal yellowing infected coconut palms. Phytopath 63:307-312
38. Ishiie T, Doi Y, Yora K and Asuyama H (1967) Suppressive effects of antibiotics of tetracycline group on symptom development in mulberry dwarf disease. Ann Phytopath Soc Japan 33:267-275
39. Johnson CG and Eden-Green SJ (1978) The search for a vector of lethal yellowing of coconuts in Jamaica: reappraisal of experiments from1962-1971. FAO Plant Protect Bull 26:137-149
40. Johnston JR (1912) The history and cause of coconut bud-rot. USDA Bureau Plant Ind. Bull 228:175
41. Leach R (1946) The unknown disease of coconut palm in Jamaica. Trop Agric 43:59-68
42. Maramorosch K (1978) Amarelecimiento letal du coqueiro: distribuicao, impacto, e implicacoes mundiais. Fitopat Brasil 3:135-148
43. Martinez AP and Roberts DA (1967) Lethal yellowing of coconut in Florida. Proc Florida State Hort Soc 80:432-436
44. Martyn EB (1945) Coconut diseases in Jamaica (I) "bronze leaf wilt" and other diseases affecting the bud of coconut. Trop Agric 22:51-59
45. McCoy RE (1972) Remission of lethal yellowing in coconut palms treated with tetracycline antibiotics. Plant Dis Reptr 56:1019-1021
46. McCoy RE (1973) Effect of various antibiotics on development of lethal yellowing in coconut palm. Proc Florida State Hortic Soc 86:503-506
47. McCoy RE (1976) Comparative epidemiology of the lethal yellowing, kaincope and cadang-cadang diseases of coconut palm. Plant Disease Reptr 60:498-502
48. McCoy RE, Miller ME and Williams DS (1980) Lethal yellowing in Texas *Phoenix* palms. Principes 24:179-180
49. McCoy RE, Howard FW, Tsai JH, Donselman HM, Thomas DL, Basham HG, Atilano RA, Eskafi FM Britt L and Collins ME (1983) Lethal yellowing of palms. Univ Florida Agric Exp Sta Tech Bull No 834

50. McCoy RE, Thomas DL and Condo JK (1976). Lethal yellowing. Why the quarantine? Florida Nurseryman 21:49-52
51. Nutman FJ and Roberts PM (1955) Lethal yellowing the "unknown disease" of coconut palms in Jamaica. Emp J Exp Agric 23:257-267
52. Ollagnier M and Weststeijn G (1961) Les maladies du cocotier aux Iles Caraibes: Comparison avec la maladie de Kaïncopé au Togo. Oléagineux 16:729-736
53. Phartasarathy MV (1974) Mycoplasmalike organisms associated with lethal yellowing disease of palms. Phytopath 64:667-674
54. Plavsic-Banjac B, Hunt P, and Maramorosch H (1972) Mycoplasma like bodies associated with lethal yellowing disease of coconut palms. Phytopath 62:298-299
55. Reinert JA (1977) Field biology and control of *Haplaxius crudus* on St. Augustine grass and Christmas palm. J Econ Entomol 70:54-56
56. Romney DH (1972) Past studies on and present status of lethal yellowing disease of coconuts. PANS 18:386-395
57. Schieber E and Hichez-Frias (1970) Lethal yellowing disease of coconut palms in the Dominican Republic. Phytopath 60:1542
58. Schuiling M (1976) A survey of insect populations on *Cocos nucifera* (Abstr.) Principes 20:67
59. Schuiling M, Johnson CG, Eden-Green SJ and Waters H (1976) Recent attempts to find a vector associated with lethal yellowing of coconut (*Cocos nucifera* L.) (Abstr.) Principes 20:65
60. Schuiling M, Kaiza DA and Mpunami A (1992) Lethal disease of coconut palm in Tanzania. II. History distribution and epidemiology. Oléagineux 47:516-522
61. Schuiling M and Mpunami A (1990) Lethal disease of coconut palm in Tanzania: Review of research up to date and preliminary results of resistance trials. In: ML Robert and D Zizumbo (Eds.) La problematica del amarillamiento letal del cocotero en México. pp. 171-183, CICY, Mérida, México
62. Schuiling M and Mpunami A (1992) Lethal disease of coconut palm in Tanzania. I. Comparison with other diseases in East Africa. Oléagineux 47:511-515
63. Seymour CP, Miller JW Roberts DA (1972) An outbreak of coconut palms lethal yellowing in Miami, Florida. Phytopath 62:788
64. Simone (1991) Lethal yellowing disease. In: AR Chase and TK Broschat (Eds.) Diseases and disorders of ornamental palms. pp. 16-20 American Phytopathology Society Press, USA
65. Thomas DL (1974) A possible link between declining palm species and lethal yellowing of coconuts. Proc Florida State Hort Soc 87:502-509
66. Thomas DL (1979) Mycoplasmalike bodies associated with lethal declines of palms in Florida. Phytopath 69:928-934
67. Thomas DL and Donselman HM (1979) Mycoplasmalike bodies and phloem degeneration associated with declining *Pandanus* in Florida. Plant Dis Reptr 63:911-916
68. Thomas DL and Norris RC (1980) The use of electron microscope for lethal yellowing diagnosis. Proc Fla State Hort Soc 93:196-199
69. Tsai JH (1977) Attempts to transmit lethal yellowing of coconut palm by the planthopper *Haplaxius crudus*. Plant Dis Reptr 61:304-307
70. Tsai JH (1979) Vector transmission of mycoplasmal agents of plant diseases. In: RF Whitcomb and JG Tully (Eds.) The mycoplasmas Vol III. pp 266-307 Academic Press, New York

71. Tsai JH and Kirsch OH (1978) Bionomics of *Haplaxius crudus* (Homoptera: Cixiidae). Environ Entomol 7:305-308
72. Waters H (1982) The absence of MLO from the salivary glands of *Myndus crudus*. In: AJ Dabek, SJ Eden-Green and H Waters (Eds.) United Kingdom ODA Lethal Yellowing of Coconut Research Scheme, Jamaica. Full Technical Report 1972-1981 pp 473-476
73. Zenner de Polania L and López AA (1977) Apuntes sobre la biología y habitos de *Haplaxius pallidus*, transmisor de la "marchitez sorpresiva" en palma africana. Rev Colomb Entomol 3(1/2):49-62

A brief history of lethal yellowing research

S.J. EDEN-GREEN
Natural Resources Institute, Central Avenue, Chatham Maritime, Chatham, Kent ME4 4TB, UK

1. Introduction

A comprehensive review of past lethal yellowing research would be largely irrelevant; there are many earlier theories, hypotheses and blind alleys that are not of concern. Several comprehensive reviews have appeared over the past 20 years [31, 34, 56, 64, 70, 71] and this paper presents a personal view of some of the landmarks in the history of lethal yellowing and related disease research, and their attempted control. This paper focuses on events prior to investigations under a European Community (EC)-funded project on the aetiology and control of lethal yellowing-type diseases (LYD) of coconut palms in Africa, in which there are six project partners. Funding for this project was formally agreed in September 1992 under the EC Science and Technology for Development Program (STD). However, some of the work started earlier with assistance from other funding agencies. The STD project was approved after proposals for research on LYD were given high priority by participants at the First International Workshop on Coconut in Africa, sponsored by BuroTrop and held in Arusha, Tanzania, in February 1991. A workshop held in Jamaica during 1992 endorsed proposals for similar activities in the Central America/Caribbean region but have yet to be translated into collaborative international projects. It is hoped that this symposium will maintain the momentum of research.

2. Significance of LYD

Lethal yellowing (LY) and related diseases (LYD) are thought to be caused by mycoplasma-like organisms (MLO), and are considered to be the most important threat to global coconut production [33]. "Caribbean" LY has existed in the northern Caribbean basin for many years and is now spreading rapidly in México

C. Oropeza et al. (eds.), Lethal Yellowing: Research and Practical Aspects, 17–33.

and Belize. In Africa, LYD are, or have been, active in Ghana (Cape St. Paul wilt), Togo (Kaïncopé), Nigeria (bronze leaf or Awka wilt), Cameroon (Kribi), Tanzania, Kenya and probably Mozambique (lethal disease).

The particular importance of LYD lies in its capacity for rapid-jump spread (attributed to insect vectors) and the genetic vulnerability of coconut palms in many regions where the diseases are active. The consequent impact of LYD on rural communities is of further concern because the coconut palm occupies a unique position as a source of food, shelter and rural income. The rapid and destructive spread of LY in the northern Caribbean is well documented [71]. In Jamaica alone, at least five million 'Jamaica Tall' palms were destroyed in the 20 years following spread of LY to the main coconut-growing areas. Similar devastation has occurred in parts of Ghana and Tanzania, but patterns of spread show greater variability. In Tanzania, the disease has killed approximately 56% of coconut palms in the southern coastal belt since 1965, but the annual rate of loss in northern districts is less than one quarter of that in the south [73].

There is particular concern in Africa because, while some coconut ecotypes are considered to be resistant to Caribbean LY [6], all are highly susceptible to LYD in East Africa [74]. Although LYD in Africa and the Caribbean show many symptomatological similarities, differences in ecotype susceptibility and epidemiology suggest that different MLOs may have evolved in each continent. The relationship between these organisms is of fundamental importance to epidemiologists, coconut breeders and those concerned with plant health and the international transfer of coconut genetic resources. However, until recently, it has not been possible to determine whether these diseases are caused by the same MLO [37]. See Harrison *et al.* (this volume) for further details.

3. Landmarks in early research on Caribbean LY

The earliest observation of LY was probably by the Marquis of Sligo, on Grand Cayman in 1834 [58]. The first scientific accounts of the disease were by Fawcett [30] and Ashby [2] in Jamaica when it was named West End bud rot thereby reflecting the distribution of the disease in the western end of the island. Johnston [48, 49] first described the disease in Cuba, where it was known simply as bud rot and may have been confused with bud rot caused by *Phytophthora*. Few reports appeared until the early 1940s when Bain [4] and Briton-Jones [7] concluded that the disease was the same as an apparently physiological disorder in Trinidad and

British Guiana (Guyana) known as bronze leaf wilt. While agreeing with the common terminology for the conditions in Jamaica and Trinidad, Martyn [58] observed that the progressive and sometimes rapid spread of the condition in Jamaica was not consistent with a physiological condition and argued that it was more likely to be an infectious disease of unknown aetiology. More detailed investigations by Leach [53] supported this conclusion and the term "unknown disease" was coined to distinguish the disease in Jamaica and the northern Caribbean, from bronze leaf wilt in Trinidad.

These and subsequent researchers were apparently unaware of the significance of independent investigations in Cuba by Bruner and Boucle [8], who concluded that the disease was most likely caused by a virus transmitted by an insect vector. The virus hypothesis was strengthened with the publication of circumstantial evidence for a viral aetiology by Nutman and Roberts [66], who first proposed the name "lethal yellowing".

The sudden spread of LY to the main coconut-growing area in the eastern end of Jamaica in 1961 provided the stimulus for two decades of intensive research in Jamaica by the Coconut Industry Board. Assistance was received from the Jamaican Ministry of Agriculture as well as international agencies including: USAID, UNDP/FAO and ODA [34]. Initial studies by Carter and co-workers [11, 12, 13, 14] concentrated on a wide range of subjects including: a thorough description of the symptomatology and epidemiology of the disease in the eastern and western ends of the island, aetiological investigations on fungi, viruses and nematodes [52], nutritional imbalances [16] and insect transmission experiments. This work provided convincing evidence that the disease was caused by an infectious agent, most probably spread by an air-borne vector. It also marked the start of many seemingly fruitless attempts to identify insect vector(s) of the disease in Jamaica.

At least two, apparently independent, outbreaks occurred in the Florida keys in the 1950s and 1960s, but the disease spread slowly and apparently died out. A notable result of these events was the claim that LY could be mechanically transmitted [68]. This claim gave rise to extensive tests in Jamaica by Grant, Grylls and Hunt, but similar results could not be repeated [32].

In addition to the abortive work on mechanical transmission, which included investigations on a soft rotting *Erwinia* sp., Grylls maintained the momentum on transmission tests and whiteflies came under strong suspicion. Tests on whiteflies were intensified by Heinze and Schuiling, the next team to arrive on the scene, who also concentrated on other Homoptera and on insecticidal control experiments [38]. Collaborative work with the Botany Department of the University of the West Indies at Mona continued through the post-graduate study of Dabek on pre- and post-symptomatic physiological and biochemical changes and the incubation period of the disease [17].

Realisation by the mid-1950s that earlier chance introductions of Malayan Dwarf palms were resistant to LY, gave impetus to the introduction and establishment of a large collection of coconut germplasm in Jamaica by the Coconut Industry Board during the early 1960s. This subsequently lead to research on the production of F_1 hybrids and establishment of a long-term breeding program.

Two important events occurred in the early-1970s: spread of the disease to mainland Florida in 1971, and identification of mycoplasma- (now mollicute-)like organisms (MLO) as the probable pathogen. The identity of the pathogen was concluded by electron-microscope observations [5, 39, 67], and subsequently supported by remission of symptoms by tetracycline antibiotics [60, 45]. These findings resulted in a massive increase in the deployment of resources for research and control of LY in Jamaica and Florida over the next decade.

4. Distribution of LY and related diseases

In addition to Jamaica, Cuba and Florida, there are reliable reports of LY affecting coconut palms from Hispaniola (Haiti and the Dominican Republic), Bahamas (New Providence Island), Mexico (Yucatán) and most recently in Belize. In Florida, more than 30 other species of palm have been affected by lethal decline, 14 of which are known to be associated with the same strain of MLO, including the Canary Island and true date palms (*Phoenix canariensis* and *P. dactylifera*). These latter species have also been affected by the same or a similar disease in Texas [64].

Elsewhere, diseases with symptoms very similar to those of LY were recognised in Nigeria (Awka wilt) more than 75 years ago [9], in Ghana (Cape St. Paul wilt) and Togo (kaïncopé disease) during the 1930s [59], in Cameroon (kribi disease) in 1937 [24], and in Tanzania and Kenya since the beginning of the century [73]. There have been reports of similar diseases in Equatorial Guinea and Mozambique.

In East Africa, on-going research on LYD (termed lethal disease or LD) commenced in Tanzania during 1980 under the auspices of the National Coconut Development Program (NCDP), with assistance from the Deutsche Gesellschaft für Technische Zusammenarbeit (GTZ) [72]. These studies have concentrated on resistance screening trails, epidemiology and identification of insect vector(s).

In Ghana during the early 1980s, seven resistance trails (totalling about 34 ha) to Cape St. Paul wilt (CSPW) were initiated under a French Government TC program, with technical management from IRHO (now CIRAD-CP) and IDEFOR, Côte d'Ivoire. Transmission studies were subsequently established following the identification of a putative vector species, *Myndus adiopodoumeensis* [57].

Apart from the trials in Ghana, research on LY in West Africa has been limited, short-term and uncoordinated. MLOs have been associated with the diseases in Ghana and Togo [20] and Cameroon [24], but not in Nigeria. Some studies on the epidemiology of these diseases have been performed [3, 24, 28, 47, 76].

Coconut diseases with some symptomatological similarities to LYD, and/or associated with MLO, have also been reported from Asia. These include diseases in peninsular Malaysia, Sarawak, Indonesia (Sumatra and Natuna Islands), Sri Lanka (leaf scorch decline) and India (root wilt and tatipaka diseases).

5. Aspects of recent research

5.1 Host range and symptomatology

The diversity of palms and closely-related species that have been introduced and established as ornamentals in southern Florida has provided unique data on the host range of Caribbean LY. Although differing in symptomatology, the similarities in epidemiology, association with MLO and, in some cases, remission of symptoms with tetracyclines, suggest that these "lethal declines" were caused by the LY agent. To date, more than 30 other palm species, and the closely related *Pandanus utilis*, have been affected by the disease. At least for 14 species, DNA probe analysis has shown that the same DNA strand is associated with all of these diseases [32] (Harrison *et al.*, this volume). In Florida, none of the few palms species native to region that have been exposed to the disease are susceptible. A similar pattern is found in Jamaica. Few of the susceptible palm species have been exposed to the disease in Jamaica but disease symptoms, accompanied by MLO, have been confirmed in *Pritchardia pacifica* and *Veitchia merrillii*. Systematic studies of the susceptibility of different palm species may provide a useful additional means to characterise different strains of the pathogen and should be considered as an adjunct to coconut germplasm resistance screening trials. In Ghana, several *Veitchia merrillii* planted alongside a coconut germplasm trial in the LYD-affected Cape Three Points region have hitherto remained unaffected by the disease.

5.2 Epidemiology

In the absence of an experimental inoculation technique, information on the probable latent period of disease in mature palms has been obtained from observations on the appearance of secondary disease spread around isolated primary foci and, in young palms, by moving palms planted in drums in and out of a disease area [19, 39]. Such studies indicate a latent period of 3-9 months in young palms and at least 7-15 months in mature ones [70]. Other features of LY epidemiology are incompletely understood. Carter and Suah [14] presented preliminary evidence that the incidence of the first appearance of symptoms was higher in the cooler autumn months. It is not known whether this reflected a corresponding earlier increase in the incidence of infection, or simply a more favourable time for symptom expression. Similar observations have recently been recorded in Ghana (Philippe, pers. comm.). Experimental evidence for variation in optimum infection periods has been difficult to obtain, but the so-called transfer experiment in Jamaica suggested that some seasons are more favourable than others [19].

Quantitative analyses have provided interesting comparisons of the rates of spread of LY and LYD in several regions and have highlighted the more variable, and usually slower, rates of spread of the diseases in Africa [61, 73, 76]. When calculating rates of absolute spread of the disease, it is important to correct for the rate of removals of sources of inoculum (diseased palms), either by the disease itself, or by cutting-out operations. However, we have little information on whether, and for how long, palms serve as a source of infection before symptoms call attention to their removal. The early detection of such palms, before they become a source of infection, could be a major advance in controlling spread of the disease.

Another poorly understood area is the factors influencing infection, especially since there is difficulty in transmitting the disease by the insect species most likely to be vectors. Does this reflect some fundamental difficulty or peculiarity in the way MLOs infect palms? For instance, are vector feeding sites or temperatures critical? Is it significant that the only reliable transmissions until now, have been in large cages or screen-houses, and if so, why?

5.3 Aetiology

The identification of MLOs as the probable causal agents of LY in 1971 raised hopes of rapid progress on the disease. At that time, there were numerous claims of successful axenic culture of plant MLOs and, had this proved possible, it would certainly have allowed for the development of a host of different diagnostic and transmission techniques. However, no claims of axenic culture could be substantiated (with the exception of those agents which proved to be spiroplasmas), but not before large amounts of time had been devoted to culturing them. There were some useful spin-offs however: plant phloem sap was found to be a suitable growth medium for cultivable mycoplasmas and spiroplasmas, although there was no evidence that it supported growth of the LY agent *in vitro*. Another interesting finding was the isolation of *Acholeplasma* spp. from diseased palms [26]. This generated considerable excitement at the time, but it was subsequently concluded that these organisms were saprophytes and not involved in the aetiology of the disease [27]. Ironically, these agents may have had greater relevance than we thought; recent work on rDNA gene sequence comparisons between plant MLOs and cultivable mollicutes, has suggested that acholeplasmas, including those isolated from palms, are the most closely related taxonomic group of organisms to plant MLOs [78].

Ultrastructural studies on the three dimensional morphology of MLO in coconut roots showed that these organisms were more complex than had been previously assumed [79]. Individual bodies could be several microns long and frequently showed complex branching, or they traversed sieve plates. Intuitively, it seemed unlikely that such complex and fragile structures would survive the mechanical disruption and pressure changes that are likely to accompany insect feeding on the phloem. Another factor mitigating against vector acquisition appeared to be the relatively low numbers of MLO in plant tissues, particularly in coconut, and their apparent concentration in photosynthetic sink areas, rather than the mature leaves which are the preferred feeding sites for most putative vectors. All of this research was performed by cumbersome and time-consuming electron microscope procedures and we can perhaps look forward to a new era in which rapid and sensitive DNA diagnostic techniques will be used on the problem.

5.4 Physiology and biochemistry

It has long been observed that the most vigorous palms appear to be the most susceptible to the disease, although it has not been determined whether this reflects factors predisposing the palm to infection by MLO, or simply making it more attractive to the vector. Carter *et al.* [13] reported that growth of the emerging fronds slowed down and ceased soon after the appearance of symptoms. However, more subtle increases in the rate of frond elongation preceding the appearance of symptoms were noted by Dabek [18] and Eden-Green and Dabek [25].

No consistent differences have been reported between diseased and healthy palms in terms of mineral content, phloem sap composition or the activities of common enzymes in phloem sap [77]. Biochemical changes preceding disease development have been observed; notably alterations in catalase activity and polyphenol content [18]. This researcher also demonstrated changes in leaf senescence and the re-greening of detached yellow pinnae from diseased palms by certain exogenously applied cations and gibberellic acid [20]. These observations were interpreted as evidence for a hormonal imbalance affecting protein metabolism in affected palms.

Antigenic differences between phloem sap from diseased and healthy palms were described by Charudattan and McCoy [15]. They also reported the presence of a common disease-specific antigen in both coconut and *Veitchia*. The nature of this antigen was not determined, but subsequent electrophoretic studies identified a characteristic disease-specific protein in *Veitchia* which increased in proportion to the stage of disease. However, small quantities of an analogous protein were also detected in healthy *Veitchia* subjected to water stress and a similar protein could not be detected in coconut [64].

Perhaps the most striking physiological changes induced by LY in coconut palms, concern the water relations of the palm, first noted in simple dye distribution experiments by Carter [11] and by Dabek [17]. Examining xylem water potential with a pressure bomb, it was found that diurnal fluctuations in water potential were markedly reduced or absent in diseased palms, and pressures were seldom less than -4 bar, in contrast to those in healthy palms which often dropped below -10 bar during the hottest period of the day [65]. The disease thus exhibited the opposite characteristics to a vascular wilt; palms were unable to transpire. Subsequent studies confirmed that an irreversible stomatal closure occurred at an early, and sometimes pre-symptomatic stage [29]. This occurrence may account for the loss of "foliar sheen" observed by Carter and co-workers some 30 years ago. See Oropeza *et al.* (this volume) for further details.

5.5 Disease transmission

Attempts to transmit LY have probably absorbed more man-years than any other aspect of research on the disease, and continues to be a key to future progress. This is particularly true in Africa where there are few clues as to the identity of the vectors. Prior to the identification of MLO, all of the major taxa of known vectors of plant disease were tested. Although largely ignored at the time, it is probable that transmissions were achieved in the large cages used for introductions of mixed Homopteran species or fulgorids by both Carter and his team and in early tests by Grylls, continued by Heinz and Schuiling. This work was reviewed by Johnson and Eden-Green [46]. It is perhaps surprising, in view of these earlier "successes", that the subsequent tests which concentrated on *Myndus crudus* and other palm-associated Homoptera, should have been uniformly unsuccessful in Jamaica.

Even in Florida, it has taken very large numbers of insects to demonstrate transmission by *Myndus* [44]. One suspects that, were it not for the strong circumstantial evidence that *M. crudus* is the main or only vector (its consistent association with palms, and the correlation of disease distribution with the distribution of its grass breeding-hosts), then there would still be strong doubts about whether it is the only, or major vector. With the DNA techniques that are now available for detecting MLO, we shall soon be hearing more about the presence of MLO in *Myndus* and other species (Harrison *et al.*, this volume). Although these results may call attention to likely vector species, transmission tests will still be necessary to confirm that they can infect palms.

The identity of vector species in Africa presents particular difficulties. *Myndus,* or closely related genera, have not been found in Tanzania where tests on the predominantly derbid planthopper fauna have produced two doubtful transmissions from over one million insects tested. As in other regions, few of the suspected vector species breed on palms and only one of them, the delphacid *Nesodryas antiope*, was considered to be sufficiently wide-spread to be a likely vector.

Myndus adiopodoumeensis occurs in parts of West Africa and has been tested as a vector in Ghana for over a year. So far, no transmissions have been noted after over 60 000 insects were introduced to test palms in very large (8 m tall) cages. Again, however, it is unlikely that this species can be solely responsible for transmission of the disease, as it has not been detected in parts of the western region of Ghana where there has recently been active disease spread. Recent surveys have shown that a species of the Typhlocibinae, *Nzinga* sp. (a small leafhopper), breeds on coconut foliage and is widely distributed on coconut palms in Ghana. This species is currently under investigation.

If spread of the disease is from palm-to-palm, as epidemiological evidence suggests, and there are no alternative hosts of epidemiological importance, then the life-cycle of the vector(s) must allow time for the latent period between acquisition of MLO from a diseased palm and the time when the insect is able to transmit. This period is rarely less than two weeks in other mycoplasma diseases. Species that breed on palm (acquiring MLO as nymphs and transmitting them as adults) would have an obvious advantage as vectors. For *Myndus,* Derbidae and most of other species that do not breed on coconut, it must be assumed that both acquisition and transmission would occur in the adult stage. The poor survival of adult insects is a recurrent problem in experimental transmission tests and may be a limiting factor in the wild. Perhaps only a small proportion of the vector population survive long enough to become capable of transmitting the disease.

5.6 Resistance

The resistance of the 'Malayan Dwarf' ecotype to LY was first recognised in the mid-1950s [66]. Seed of this ecotype had been introduced to Jamaica on at least two previous occasions, before there was any interest in it as a possible source of resistance to the disease. Other sources of resistance were subsequently observed in a few existing plantings of 'Panama Tall' and in a population of the progeny of 'Malayan Dwarf' x 'Fiji Dwarf' ('Niu Leka') F_1 hybrids. More extensive data were obtained from a series of germplasm trials planted in the mid-1960s from collections carried out with assistance from The Overseas Development Administration (ODA) and the Food and Agriculture Organisation (FAO) [80, 81].

Results showed further promising sources of resistance, notably in the 'Sri Lanka Dwarf', 'Indian Dwarf' and 'King Coconut' [6]. Intermediate resistance was observed in a few of the tall ecotypes. Development of techniques for the mass controlled pollination allowed the large-scale introduction of F_1 hybrids combining the high resistance of the 'Malayan Dwarf' with the more attractive agronomic characteristics of the more susceptible tall parents. The 'Maypan' hybrid was successfully introduced and adopted in Jamaica [62] and elsewhere in the Caribbean region.

In Tanzania, more than thirty different ecotypes and hybrids have been screened, including those identified as highly resistant to LY in Jamaica. However, none showed acceptable resistance to lethal disease under conditions of high disease intensity [72]. Possible sources of resistance may lie in sub-populations or in land-races of the local 'East Africa Tall' in Tanzania. These are now under investigation. Early results from the seven germplasm trials planted about 10 years ago in the southern and western regions of Ghana show yet a different pattern of susceptibility. The 'Sri Lanka Green Dwarf', 'Equatorial Green Dwarf', 'Malayan Yellow Dwarf' (but not the red form) and 'Vanuatu Tall' have shown few losses to Cape St. Paul wilt in the two trials that have been affected by the disease. In contrast, the local 'West Africa Tall' have been almost completely destroyed by the disease.

Although the differences in ecotype susceptibility in different regions may be due to environmental or biotic factors mediating resistance (*e.g.* temperature or vector preferences), there are also different strains of LYD MLO in Africa and the Caribbean (Harrison *et al.*, this volume; Jones *et al.*, this volume). Little is known about the nature of resistance to LYD. A correlation between susceptibility and free arginine content has been noted in a limited range of coconut ecotypes and palm species, and it has been suggested that this might be related to the ability of many mollicutes to metabolise arginine [62]. The intermediate susceptibility of dwarf x tall hybrids suggests that resistance may be under polygenic control. However, resistance is rarely absolute and those resistant dwarfs that do contract the disease exhibit the same range and severity of symptoms as susceptible ecotypes. Therefore, resistance may operate at the stage of infection, rather than through some physiological or biochemical property inimical to multiplication of the pathogen.

5.7 Control

Where it has been found, resistance has provided a highly effective means of coping with LYD, as the Jamaican experience shows (Been, this volume). The early introduction, screening and establishment of germplasm likely to resist the disease should thus be the first option in disease control [34]. However, true-breeding ecotypes may not combine the most favourable agronomic characteristics with acceptable resistance to the disease, and production of hybrids may be a complementary activity. In most areas, the resistance of the Malayan Dwarf to Caribbean LY has been remarkably stable, even under intense disease pressure. However, occasional but so far localised, outbreaks of LY in Malayan Dwarf palms in Jamaica and Florida [42] provide a worrying reminder of the risk of selecting strains of the pathogen that can overcome this source of resistance.
Quarantine and sanitation measures against LYD have a controversial history. Although legislation against movement of planting materials of coconut and other susceptible palms has been enacted in Jamaica and Florida, there is little evidence that such controls have any impact on containing the disease. Transmission through seed is so far unknown in MLO diseases and observations on seedlings grown from seed collected from LY diseased palms have produced no evidence that this occurs in coconut [71]. Few, however, wish to take the risk of proving the exception to the rule. Perhaps DNA detection techniques will elaborate on this issue.

Control campaigns have involved the elimination of diseased palms and/or application of insecticides or tetracycline antibiotics around sources of infection. When continuously applied, tetracycline treatments proved an effective short-term prophylactic measure for the high value ornamental and amenity plantings in Florida [63]. However, antibiotics are unlikely to find application in low-input agricultural systems. There have been few convincing or statistically significant examples of success using cutting-out or insecticide treatments [38]. Even where intensive insecticide treatments have been shown to reduce the rate of spread of disease, as in sidewalk plantings of ornamental palms in Florida [43], it is difficult to conceive how the necessary resources could be deployed in the low-input and often inaccessible cultivations of developing countries. However, such treatments may have a role in controlling new isolated outbreaks or introductions of the disease. Such treatments are thought to have been successful in recent outbreaks in the Tanzanian offshore island of Mafia (Mpunami pers. comm.). There are also suggestions that palm removal and insecticide treatments may have slowed the rate of spread in parts of Ghana (Arkhurst pers. comm.). Alternative strategies are to reduce vector populations by controlling their breeding hosts or substituting a ground cover of non-host species [40, 41], or by exploiting a potential biological control agent such as *Hirsuitella* [10]. It should be observed, however, that control of disease spread probably requires a greater reduction in populations of an insect vector than would be necessary for an insect pest that causes damage in its own right. Answers to these problems can only be obtained through research, but trials are difficult and demand large resources and long-term commitment to implementation and monitoring.

6. Conclusions- recent developments

A recurrent theme of previous LYD control campaigns and research projects has been the problem of detection and diagnosis of MLO. This has been the key not just for rapid, reliable and sensitive detection of these non-cultivable agents in plants and insect vectors, but also to establishing the relationships between the diseases in different regions. Until recently, progress in this area was beyond the reach of available technology. Disease-specific antibodies, nucleic acid probes [22, 50] and selective DNA amplification with MLO-specific PCR primers [1, 23] are now available to determine relationships between MLOs and to monitor their distribution and ecology [51, 54, 55]. Progress has been made in developing techniques to detect MLO associated with LYD in the Caribbean region [36] and East Africa [69]. These techniques are now being used to investigate the relatedness of MLO from Africa and the Caribbean (Harrison *et al.*, this volume; Jones *et al.*, this volume).

These technologies have opened exciting new prospects for the application of practical diagnostic tests that may be used for understanding the ecology, epidemiology and relatedness of LYD in different regions. Using these techniques, MLO can now be routinely detected in samples from both East and West Africa by PCR. Different MLOs can be identified and distinguished by comparing restriction fragment length polymorphisms (RFLPs) of the PCR products and the picture that is emerging suggests that there are different strains of LYD MLO in those regions.

The long-term solution to LYD most likely lies in the identification and manipulation of resistance to the disease. Fortunately, or perhaps fortuitously, sources of resistance can be found to at least some strains of the pathogen. A better characterisation, and ultimately classification of strains of the MLO may allow the prediction of susceptibility of palm genotypes to disease in different regions.

However, for the foreseeable future, such predictions will rely on the hard-won data from field trials performed under often remote and difficult conditions. Finally, I acknowledge the researchers, past and present, who have had the foresight and perseverance to collect and screen coconut germplasm for resistance to LY.

References

1. Aherns U and Seemüller E (1992) Detection of DNA of plant mycoplasmalike organisms by a polymerase chain reaction that amplifies a sequence of the 16S rRNA gene. Phytopath 82:828-832
2. Ashby SF (1915) Notes on diseases of cultivated crops observed 1913-1914. Bull Dept Agric Jamaica 2:299-311
3. Bachy A and Hoestra H (1958) Contribution a l'etude de la maladie de Kaïncopé du cocotier au Togo. Oléagineux 13:721-730
4. Bain FM (1940) Report on the coconut growing areas of Jamaica. Dept Sci Agric Jamaica Bull No 2
5. Beakbane AB, Slater CHW and Posnette AF (1972) Mycoplasmas in the phloem of coconut, *Cocos nucifera* L., with lethal yellowing disease. J Hort Sci 47:265
6. Been BO (1981) Observations on field resistance to lethal yellowing in Jamaica. Oléagineux 36:9-11
7. Briton-Jones HR (1940) The disease of the coconut palm. London: Ballière, Tindall and Cox
8. Bruner SC and Boucle L (1943) La enfermada conocida por "pudricion del cogollo" del cocotero en Cuba. Rev Agric Cuba 26:132-141
9. Bull RA (1955) Bronze leaf wilt of coconut palms in Nigeria. J West Afric Inst Oil Palm Res 3:70-72
10. Carillo HH and Piña JR (1990) Situation actual del amarillamiento letal en el sureste de México. In: Robert ML and Zizumbo D, eds. La problematica del amarillamiento letal del cocotero en México, pp 69-93. Mérida: Centro de Investigación Científica de Yucatán
11. Carter W (ed.) (1965) FAO lethal yellowing disease project: a compilation of data derived during the FAO mission from June, 1962, to June, 1965, Coconut Industry Board, Kingston, Jamaica
12. Carter W (1966) Susceptibility of coconut palm to lethal yellowing disease. Nature 212:320
13. Carter W, Latta RK and Suah JRR (1965) The symptoms of lethal yellowing disease of coconut palms. FAO Plant Protect Bull 13:49-55
14. Carter W and Suah JRR (1965) Studies on the spread of lethal yellowing disease of the coconut palm. FAO Plant Protect Bull 12:73-78
15. Charudattan R and McCoy RE (1975) Antigenic difference in phloem exudates of healthy and lethal yellowing diseased coconut palms. Proc Am Phytopath Soc 2:71
16. Chen RA (1966) Nutritional aspects of lethal yellowing in coconuts. Trop Agric 43:211-218
17. Dabek AJ (1973) Some physiological and anatomical aspects of the lethal yellowing disease of coconut. PhD thesis, University of the West Indies
18. Dabek AJ (1974) Biochemistry of coconut palms affected with the lethal yellowing disease in Jamaica. Phytopath Z 81:346-353
19. Dabek AJ (1975) The incubation period, rate of transmission and effect on growth of coconut lethal yellowing disease in Jamaica. Phytopath Z 84:1-9
20. Dabek AJ and Hunt P (1976) Biochemistry of leaf senescence in coconut lethal yellowing, a disease associated with mycoplasma-like organisms. Trop Agric 53:115-123
21. Dabek AJ, Johnson CG and Harries HC (1976) Mycoplasma-like organisms associated with kaincope and Cape St. Paul wilt diseases of coconut palms in West Africa. PANS 22:354-358
22. Davis MJ, Tsai JH, Cox RL, McDaniel LL and Harrison NA (1988) Cloning of chromosomal and extrachromosomal DNA of the mycoplasmalike organism that causes maize bushy stunt disease. Mol Plant-Microbe Interact 1:295-302

23. Deng S and Hiruke C (1991) Genetic relatedness between two nonculturable mycoplasmalike organisms revealed by nucleic acid hybridisation and polymerase chain reaction. Phytopath 81:1475-1479
24. Dollet M, Gianotti J, Renard JL and Ghosh SK (1977) Etude d'un jaunissement lethal des cocotiers au Cameroun: la maladie de Kribi. Observations d'organismes de type mycoplasmes. Oléagineux 32:317-322
25. Eden-Green SJ and Dabek AJ (1975) Effects of lethal yellowing on growth of coconut palms. Principes 20:61-62.
26. Eden-Green SJ and Tully JG (1979) Isolation of *Acholeplasma* spp. from coconut palms affected by lethal yellowing disease in Jamaica. Curr Microbiol 2:311-316
27. Eden-Green SJ, Markham PG, Townsend R, Archer DB, Clark MF and Tully JG (1984) Plant pathogenicity and transmission tests with *Acholeplasma* spp. isolated from coconut palms in Jamaica. Ann Appl Biol 106:439-449
28. Ekpo EN and Ojomo EE (1990) The spread of lethal coconut diseases in West Africa: incidence of Awka disease (or bronze leaf wilt) in the Ishan area of Bendel State of Nigeria. Principes 34:143-146
29. Eskafi FM, Basham HG and McCoy RE (1986) Decreased water transport in lethal yellowing-diseased coconut palms. Trop Agric 63:225-228
30. Fawcett W (1891) Report on the coconut disease at Montego Bay. Bull Bot Dept Jamaica 23:2
31. Grylls NE and Hunt P (1971) A review of the study of the aetiology of coconut lethal yellowing disease. Oléagineux 26:311-315
32. Grylls NE and Hunt P (1971) Studies on the aetiology of coconut lethal yellowing in Jamaica, by mechanical and bacteria inoculations and by insect vectors. Oléagineux 26:453-549
33. Harries HC (1978) Lethal yellowing disease of coconuts in global perspective. Philipp J Coconut Stud 3:1-4
34. Harries HC (1990) Lethal yellowing disease of coconuts in Jamaica: history, strategy and results. In: Robert ML and Zizumbo D, eds. La problematica del amarillamiento letal del cocotero en México, pp 159-169. Mérida: Centro de Investigación Científica de Yucatán
35. Harries HC and Romney DH (1974) Maypan: an F_1 hybrid coconut variety for commercial production in Jamaica. World Crops 26:110-111
36. Harrison NA, Bourne CM, Cox RL, Tsai JH and Richardson PA (1992) DNA probes for detection of mycoplasmalike organisms associated with lethal yellowing disease of palms in Florida. Phytopath 82:216-244
37. Harrison NA, Richardson PA, Jones P, Tymon AM, Eden-Green SJ and Mpunami AA (1994) Comparative investigation of mycoplasma-like organisms (MLOs) associated with Caribbean and African coconut lethal decline disease by DNA dot hybridisation and PCR assays. Plant Dis 78:507-511
38. Heinze KG, Petzold H and Marwitz R (1972) Beitrag zür atiologie der todlichen vergilbung der kokospalme. Phytopath Z 74:230-237
39. Heinze KG, Schuiling M and Romney DH (1972) The possible cause of lethal yellowing disease of coconut. FAO Plant Protect Bull 20:58-68
40. Howard FW (1980) Population densities of *Myndus crudus* van Duzee (Homoptera: Cixiidae) in relation to coconut lethal yellowing distribution in Florida. Principes 24:174-178
41. Howard FW (1990) Evaluation of grasses for cultural control of *Myndus crudus*, a vector of lethal yellowing of palms. Entomol Exp Appl 56:131-137
42. Howard FW, Atilano R, Barrant CI, Harrison NA, Theobald WF and Williams DS (1987) Unusually high lethal yellowing disease incidence in Malayan Dwarf coconut palms on localised sites in Jamaica and Florida. J Plantation Crops 15:86-100
43. Howard FW and McCoy RE (1980) Reduction in spread of mycoplasma-like organsim associated lethal decline of the palm *Veitchia merrillis* by the use of insecticides. J Econ Entomol 73:268-270
44. Howard FW, Norris RC and Thomas DL (1983) Evidence of transmission of palm lethal yellowing agent by a planthopper, *Myndus crudus* (Homoptera: cixiidae). Trop Agric 60:168-171

45. Hunt P, Dabek AJ and Schuiling M (1974) Remission of symptoms following tetracycline treatment of lethal yellowing infected coconut palms. Phytopath 64:307-312
46. Johnson CG and Eden-Green SJ (1978) The search for a vector of lethal yellowing of coconuts in Jamaica: reappraisal of experiments from 1962-1971. FAO Plant Protect Bull 9:125-129
47. Johnson CG and Harries HC (1976) A survey of Cape St. Paul wilt in West Africa. Ghana J Agric Sci 9:125-129
48. Johnston JR (1909) The bud rot of the coconut palm. USDA Bur Plant Indust Circ No 36
49. Johnston JR (1912) The history and cause of the coconut bud rot. USDA Bur Plant Indust Bull No 228
50. Kirkpatrick BC, Stenger DC, Morris TJ and Purcell AH (1987) Cloning and detection of DNA from a nonculturable plant pathogenic mycoplasma-like organism. Science 238:197-200
51. Kuske CR and Kirkpatrick BC (1992) Distribution and multiplication of western aster yellows mycoplasma-like organisms in *Catharanthus roseus* as determined by DNA hybridisation analysis. Phytopath 82:457-462
52. Latta R (1966) Attempts to relate nematodes to lethal yellowing disease of the coconut palm in Jamaica. Trop Agric 43:59-68
53. Leach R (1946) The unknown disease of the coconut palm in Jamaica. Trop Agric 23:50-60
54. Lee I-M and Davis RE (1988) Detection and investigation of genetic relatedness among aster yellows and other mycoplasmalike organisms by using cloned DNA and RNA probes. Molec Plant-Microbe Interact 1:303-310
55. Lee I-M, Davis RE and DeWitt ND (1990) Nonradioactive screening method for isolation of disease-specific probes to diagnose plant diseases causes by mycoplasmalike organisms. Appl Environ Microbiol 56:1471-1475
56. Maramorosch K and Hunt P (1981) Lethal yellowing disease of coconut and other palms. In: Raychaudhuri SP and Maramorosch K, eds. Mycoplasma Diseases of Trees and Shrubs, pp 185-210. New York: Academic Press
57. Mariau D (1990) Contribucion del IRHO a la investigacion sombre el amarillamiento letal del cocotero en México. In: Robert ML and Zizumbo D, eds. La problematica del amarillamiento letal del cocotero en México, pp 185-187. Mérida: Centro de Investigación Científica de Yucatán
58. Martyn EB (1945) Coconut diseases of Jamaica. (I). Bronze leaf wilt and other diseases affecting the bud of coconuts. (II). Diseases affecting the leaves, crown and stem of coconuts. Trop Agric 22:51-59,69-76
59. Martyn EB (1955) Diseases of coconuts. Trop Agric 32:162-169
60. McCoy RE (1972) Remission of lethal yellowing in coconut palm treated with tetracycline antibiotics. Plant Dis Reptr 56:1019-1021
61. McCoy RE (1976) Comparative epidemiology of the lethal yellowing, kaincope and cadang-cadang diseases of cococnut palm. Plant Dis Reptr 60:498-502
62. McCoy RE (1984) Presence of free arginine associated with susceptibility to lethal yellowing of palms. Israel J Med 20:1061-1081
63. McCoy RE, Carroll VJ, Poucher CP and Gwin GH (1976) Field control of lethal yellowing with oxytetracycline hydrochloride. Phytopath 66:1148-1150
64. McCoy RE, Howard FW, Tsai JA, Donselman HM, Thomas DL, Basham HG, Atilano RA, Eskafi FM, Britt L and Collins ME (1983) Lethal yellowing of palms. Univ Florida Agric Exp Sta Bull No 834
65. McDonough J and Zimmerman MH (1979) Effect of lethal yellowing on xylem pressure in coconut palms. Principes 23:132-137
66. Nutman FJ and Roberts PM (1955) Lethal yellowing: the "unknown disease" of coconut palms in Jamaica. Emp J Agric 23:257-267
67. Plavsic-Banjac B, Hunt P and Maramorosch K (1972) Mycoplasma-like bodies associated with lethal yellowing disease of coconut palms. Phytopath 58:298-299
68. Price WC, Martinez AP and Roberts DA (1968) Reproduction of coconut lethal yellowing syndrome by mechanical inoculation of young seedlings. Phytopath 58:593-596

69. Rohde W, Kullaya A, Mpunami A and Becker D (1993) Rapid and sensitive diagnosis of mycoplasmalike organisms associated with lethal disease of coconut palms by a specially primed polymerase chain reaction for the amplification of 16S rDNA. Oléagineux 48:319-322
70. Romney DH (1972) Past studies and present status of lethal yellowing disease of coconuts. Pest Artic News Summ 18:386-395
71. Romney DH (1983) Brief review of coconut lethal yellowing. Indian Coconut J 13:1-8
72. Schuiling M and Mpunami A (1992) Lethal disease of coconut palm in Tanzania I. Comparison with other diseases in East Africa. Oléagineux 47:511-515
73. Schuiling M, Kaiza DA and Mpunami A (1992) Lethal disease of coconut palm in Tanzania II. History, distribution and epidemiology. Oléagineux 47:516-521
74. Schuiling M, Mpunami A, Kaiza DA and Harries HC (1992) Lethal disease of coconut palm in Tanzania III. Low resistance of imported germplasm. Oléagineux 47:693-697
75. Schuiling M, Neinhaus F and Kaiza DA (1981) The syndrome in coconut palms affected by lethal disease in Tanzania. Z Pflznkrank Pflznschutz 85:665-677
76. Steiner KG (1976) Epidemiology of Kaïncopé disease of coconut palms inTogo. Plant Disease Reptr 60:613-617
77. Stemmer WPC, Archer DB, Daniels MJ, Davies AMC and Eden-Green SJ (1982) Effects of lethal yellowing on composition of phloem sap from coconut palms in Jamaica. Phytopath 72:672-675
78. Toth KE, Harrison NE and Sears BB (1994) Phylogenetic relationships among members of the class Mollicutes deduced from *rps3* gene sequences. Int J Syst Bacteriol 44:119-124
79. Waters H and Hunt P (1980) The *in vivo* three-dimensional form of a plant mycoplasma-like organism revealed by the analysis of serial ultrathin sections. J Gen Microbiol 116:111-131
80. Whitehead RA (1966) Report of sample survey and collection of germplasm in the Pacific Islands. 30 May to 5th September, 1964. ODM Research Publication No 16. London: HMSO
81. Whitehead RA (1968) Selecting and breeding coconut palms resistant to lethal yellowing disease. Euphytica 17:81-101

Mycoplasmalike organisms as pathogens of coconut palms

PHILIP JONES, ANATOLIA MPUNAMI and ANNA TYMON
Plant Pathology Department, Rothamsted Experimental Station, Harpenden Herts AL5 2JQ, UK

1. Introduction

Lethal diseases of coconut palms have been reported for many years from all coconut-producing areas. As investigative methods have improved, the pathogens associated with many of these diseases have been identified and Koch's Postulates fulfilled. However, for one group of pathogens, the plant mycoplasmalike organisms (MLOs), failure to isolate the pathogens and to study them *in vitro* has seriously restricted knowledge of them.

Until 1991, most knowledge of MLOs was based on comparative studies of the culturable mycoplasmas of clinical or veterinary importance as well as *Spiroplasma citri*, the cause of citrus stubborn disease. Diseases caused by MLOs were characterised on symptomatology, host range and vectors, electron microscopy of MLO structure in the plant sieve tubes, antibiotic sensitivity and rudimentary serological studies. Such studies gave basic information but were of little help in characterising the organisms that affect over 300 plant species [26].

Studies of the phylogeny of plant MLOs with true members of the *Mycoplasmatales* were only possible after recombinant DNA technology became available. The cloning of MLO DNA fragments from the insect vector of Western-X disease by Kirkpatrick *et al.* [20] signalled the start of a new age in the study of plant diseases caused by MLOs. In coconut palms there are at least twelve diseases that are either known to be associated with MLOs or are still of uncertain aetiology.

MLOs are cell wall-less prokaryotes which are members of the class Mollicutes. They are comprised of a single plasma-membrane and have a variable shape but often appear rounded. The internal structure of an MLO consists of ribosome-like

C. Oropeza et al. (eds.), Lethal Yellowing: Research and Practical Aspects, 35–42.

granules and strands of DNA. All electron microscopic studies of MLOs in plant sieve tubes show essentially the same information and it has been impossible to differentiate between MLOs on the basis of morphology and ultrastructure. In spite of this, electron microscopy became an essential step in proving the association between an MLO and a disease of unknown aetiology [8, 26]. Since electron microscopy was expensive in both time and resources, optical microscopy has also been used to detect MLOs in plants. Both chromatic (dienes stain) and fluorescent (DAPI-4, 6-diamidino-2 penylindole) stains have been used to show infection of phloem by MLOs [10, 34, 39]. However, optical microscopy again provides no information on the identity of the MLO.

2. Symptoms and spread

Symptoms exhibited by plants infected with MLOs include virescence, phyllody, proliferation, phloem necrosis, yellowing of foliage and decline of vigour leading to death of the plant. There have been a number of attempts by researchers to classify MLO diseases on the basis of symptoms. Chiykowski and Sinha [7], Shiomi and Sugiura [35] and Kirkpatrick [19] have produced classifications in which two or more groups of plant MLOs were postulated. The categories proposed by Kirkpatrick were: (i) the virescence inducing MLOs and (ii) the decline diseases which include: western-X disease, apple proliferation, elm yellows and lethal yellowing of coconut palms.

Both plants and insects act as hosts for MLOs where they induce systemic infections. Experimentally determined host ranges are considered to be of limited value; concluded from research involving the transmission of the western aster yellows MLO by the vector *Aphrodes bicintus*, and the clover phyllody MLO by the vector *Macrosteles frascifrons* [7]. They concluded that differences in host range could be accounted for by differences in vector feeding, rather than true differences in pathogenicity. This problem becomes even more acute when dealing with experimental hosts such as coconut palms (*i.e.* large trees) rather than small herbaceous annuals or perennials in screen-houses.

Definite associations between coconut palms and MLOs have been established for lethal yellowing in the Caribbean [29], Cape St. Paul wilt, kaïncopé, kribi and Awka wilt in West Africa [9, 12, 15], and lethal disease (LD) in Tanzania and Kenya [27] (Tymon, Mpunami and Jones unpubl.) and tentative associations exist for coconut stem necrosis in Malaysia [41] and coconut (root) wilt in India [40].

Symptomatologically, the MLO diseases of coconut palms are essentially similar. In bearing palms, the first symptoms are usually a premature fruit fall, closely followed by the appearance of blackened newly-opened inflorescences. These symptoms are then followed by a progressive discoloration and shedding of foliage, starting from the oldest fronds. The spear leaf becomes necrotic, with the necrosis spreading down to the tissues surrounding the growing point where a foul-smelling internal rot develops. The growing point may remain intact until most of the foliage is affected, but eventually the crown of the palm rots and falls off within 3-6 months after appearance of the first symptoms. The coloration of the affected fronds is a varietal characteristic and is particularly noticeable in 'Jamaica Tall'. In this ecotype, the fronds turn yellow; other ecotypes may show bronze or brown coloration.

When compared with lethal yellowing, some differences become apparent between coconut diseases associated with MLOs from other parts of the world. Whereas Caribbean lethal yellowing is characterised by a rapid jump spread pattern [25, 32], this type of spread is rare with Tanzanian LD [36].

Varietal resistance patterns also differ. For example, the dwarf varieties Malayan and Sri Lankan Green, which are considered to be highly resistant to Caribbean lethal yellowing [3], are very susceptible to Tanzanian LD [38]. Such differences may, however, be due to vector feeding preferences rather than varietal resistance. The vector of Caribbean lethal yellowing is considered to be *Myndus crudus* [16], but this may not be the only vector and, to date, no examples of this genus have been found in association with coconuts in Tanzania [37].

Lethal disease (LD) occurs throughout the coastal belt of mainland Tanzania, but not on the islands of Zanzibar and Pemba. Even on the mainland, the severity of the disease differs markedly among the regions affected. The disease is very severe in southern regions where some 56% of palms have been killed since 1965, but in the northern regions, less than 10% have been affected [36].

In parts of West Africa, there are also no reports of resistance to the indigenous MLO diseases of coconut. Dollet *et al.* [12] reported 'Cameroon Red Dwarf' being affected by kribi disease in Cameroon and kaïncopé in Togo. However, in recent trials in Ghana, two ecotypes which are susceptible to lethal yellowing in the Caribbean ('Panama Tall' and 'Sri Lankan Green Dwarf') have not yet been affected by Cape St. Paul wilt, which contradicts earlier reports [18].

One feature exhibited by all lethal yellowing diseases of coconuts is their rapid and destructive spread. This is well documented in the Caribbean. Similar devastation to the Caribbean LY is seen in parts of Ghana and Tanzania, but in Africa, patterns of spread are more variable [37] and losses may be more manageable. It seems likely, however, that most, if not all, of the world´s coconut palms are susceptible to some mycoplasma pathogen.

3. Distribution

Other coconut diseases occurring as erratic localised outbreaks often have symptoms similar to lethal yellowing, especially in Asia (*e.g.* Malaysia wilt and Natuna wilt [2]), but no putative pathogen has been identified. Turner *et al.* [41] described a disease they called coconut stem necrosis which caused considerable losses in young plantations of introduced Malayan Dwarf and hybrid palms in Indonesia and peninsular Malaysia. Mycoplasmalike organisms were found by electron microscopy in some palms affected by this disease. The symptoms of stem necrosis differed markedly from those of lethal yellowing or LD and may be related to coconut blast reported from West Africa [30]. In southern India, the syndrome known as Kerala wilt or coconut (root) wilt [5, 17] has been the subject of investigation and debate by pathologists for many years. The slow decline symptoms are markedly different from those of lethal yellowing and LD and factors other than an MLO pathogen may have a role in its aetiology.

4. Diagnostics

In common with other plant pathogenic MLOs, those causing lethal diseases of coconuts cannot be cultured, thus making progress in studying the pathogens slow and difficult. Serological assays are routinely used for culturable prokaryotes, but difficulties with the purification of the plant MLOs meant that first attempts at purification resulted in cross-reacting sera. It is interesting to note that in the early-1950s, Nutman and Roberts [28] attempted to produce a diagnostic antiserum for lethal yellowing using sap from diseased palms. Unfortunately the results were inconclusive. Monoclonal antibodies have been raised to a number of MLOs including: aster yellow, clover phyllody, flavescence dorée and elm yellows [4, 6, 13]. These monoclonals are useful for differentiating closely related strains of MLOs, while polyclonal antibodies can differentiate the more distantly related MLOs [22]. As far as we can ascertain, there have been no comprehensive tests of antibodies raised to plant MLOs with the coconut MLOs.

The application of molecular biological techniques to the study of plant pathogenic MLOs has opened up new avenues for researchers, but only the dedicated, or perhaps foolhardy, devote their time to MLOs in coconut palms. DNA hybridisation studies using cloned random fragments of MLO DNA extracted from plants as probes has made it possible to reliably and specifically detect MLOs in their plant and insect hosts [22]. The DNA is often prepared from plants which have high titres of MLOs such as periwinkle (*Catharanthus roseus*). Harrison *et al.* [14] were able to extract MLO DNA from infected palms by multiple cycles of ultra-centrifugation in cesium chloride-bis benzimide gradients and showed that these probes were specific for the lethal yellowing MLO.

5. Diversity

Cloning of random fragments of MLO DNA can produce probes which will hybridise to a range of other MLOs. Such probes have been used to study the phylogeny of plant MLOs. A number of plant MLO classifications have appeared during the past two years and a consistency of classification is beginning to appear [21, 23, 24].

Amplification of 16S rDNA from MLOs by the polymerase chain reaction (PCR) provides a much more sensitive detection method than any other hitherto used. However, the choice of oligonucleotide primers can affect the specificity of the test. In our laboratory, primers designed by Deng and Hiruke [11], Lee *et al.* [24], Rohde *et al.* [31] and Kirkpatrick (pers. comm.) have been used with varying success to amplify MLO DNA from coconut from East and West Africa, India and date palms (*Phoenix dactylifera)* from North Africa. The object of our research was to design primers that would differentiate the coconut MLOs from other plant MLOs and, if possible, from each other. The techniques will then be transferred to our collaborators in Africa for use in their national programs.

Using PCR and restriction enzyme digests, it can be shown that the LD MLOs of Tanzania and Kenya are similar, and may be identical but that they differ from the Cape St. Paul wilt and Awka disease MLOs. These finding may help to account for the epidemiological differences observed. Both African MLOs differ, again, from the Caribbean lethal yellowing MLO, which, in the limited tests so far done, is more closely related to the East African MLO than the West African one. MLO DNA has been amplified in three out of four samples of coconut with coconut root wilt disease, but so far attempts to differentiate this MLO from the East African MLO by restriction enzyme digests have failed. We hope that we can obtain further samples from Kerala to pursue our investigations. Similarly, MLO DNA amplified from samples of date palm have also been difficult to differentiate from that of the LD MLO.

It is important to remember that these analyses are based on PCR products amplified from the 16S rRNA gene, which represents a very small portion of the total MLO genome. As this gene is known to be highly conserved in nature, the lack of diversity is not surprising. It is possible that primers based on cloned random DNA fragments will provide a means to distinguish between these MLOs [23, 33] (Harrison pers. comm.). Cloning and sequencing of PCR products and other chromosomal DNA fragments should enable primers to be designed which are specific to each of the differing coconut MLOs. These primers can then be used in the search for vectors and possible alternate hosts of the pathogens, and will also help breeders in their search for resistant germplasm.

6. Conclusions

In conclusion, evidence is beginning to emerge that lethal diseases of coconut are not caused by a single organism, but by a group of at least three closely related strains. If the MLO proves to be carried by viable embryos, a close study of the phylogeny of the coconut MLOs may throw some light on the theories about the dissemination of coconut palms.

Acknowledgements

We thank Dr N.A. Harrison, IFAS, University of Florida, Fort Lauderdale, for helpful discussions and encouragement, and the British Council (México), the German Agency for Technical co-operation and Development (GTZ) and the European Community for financial support.

References

1. Addison EA (1978) The Cape St Paul wilt disease in Ghana; the present position. Proc 3rd ICLY Meeting. Publ Fl-78-2, p 7, Agric Res Center, University of Florida, Fort Lauderdale.
2. Anon (1981) Yearly Progress Report of Coconut Research 1980. Rome: Food and Agriculture Organisation
3. Been B (1981) Observations on field resistance to lethal yellowing in Jamaica. Oléagineaux 36:9-11
4. Boudon-Padieu B, Larrue J and Caudwell A (1989) ELISA and dot-blot detection of flavescence dorée MLO in individual leafhopper vectors during latency and inoculative state. Curr Microbiol 19:357-364
5. Butler EJ (1908) Report on coconut palm diseases in Travencore. Agric Res Inst Pusa Bull No 9
6. Chang TI and Chen TA (1991) Utilisation of monoclonal antibodies against elm yellows mycoplasmalike organisms in detection of elm yellows disease. Phytopath 81:121
7. Chiykowski LN and Sinha RC (1990) Differentiation of MLO diseases by means of symptomatology and vector transmission. Zentrallblatt Bakteriol Suppl 20:280-287
8. Cousin MT, Sharma AK and Misra S (1986) Correlation between light and electron microscopic observations and identification of mycoplasmalike organisms using consecutive 350 nm thick sections. J Phytopath 115:368-374
9. Dabek AJ, Johnson CG and Harries HC (1976) Mycoplasma-like organisms associated with kaincope and Cape St. Paul wilt diseases of coconut palms in West Africa. PANS 22:354-358
10. Deely J, Stevens WA and Fox RTV (1979) Use of dienes stains to detect plant diseases induced by mycoplasmalike organisms. Phytopath 69:1169-1171
11. Deng S and Miruki C (1991) Amplification of 16S rRNA genes from culturable and nonculturable mollicutes. J Microbiol Methods 14: 53-61
12. Dollet M, Gianotti JL, Renard JL and Ghosh SK (1977) Study of a lethal yellowing of coconuts in Cameroon: kribi disease. Observations of mycoplasmalike organisms. Oléagineux 32:317-322
13. Errampali D, Fletcher J and Sherwood JL (1989) Production of monospecific polyclonal antibodies against the aster yellows mycoplasmalike organisms (AY MLO) of Oklahoma. Phytopath 79:1137
14. Harrison NA, Bourne CM, Cox RL, Tsai JH and Richardson PA (1992) DNA probes for detection of mycoplasmalike organisms associated with lethal yellowing of palms in Florida. Phytopath 82:216-224
15. Harrison NA, Richardson PA, Jones P, Tymon AM, Eden-Green SJ and Mpunami AA (1994) Comparative investigation of mycoplasmalike organisms associated with Caribbean and African coconut lethal decline diseases by DNA hybridisation and PCR assays. Plant Dis 78:507-511
16. Howard FW, Norris RC and Thomas DL (1983) Evidence of transmission of palm lethal yellowing agent by a planthopper *Myndus crudus* (Homoptera: Cixildae). Trop Agric 60:186-171
17. Jayasankar NP and Bavankar KVA (1986) Coconut root wilt disease- past studies, present status and future strategy. Indian J Agric Sci 50:309
18. Johnson CG and Harries HC (1976) A survey of Cape St. Paul wilt of coconuts in West Africa. Ghana J Agric Sci 9:125-129
19. Kirkpatrick BC (1989) Strategies for characterising plant pathogen mycoplasmalike organisms and their effects on plants. In T Hosuge and EW Nester, eds. Plant-Microbe Interactions, Molecular and Genetic Perspective, volume 3, pp 241-293. New York: McGraw-Hill
20. Kirkpatrick BC, Stenger DC, Morris TJ and Purcell AH (1987) Cloning and detection of DNA from a non-culturable plant pathogenic mycoplasmalike organism. Science 238:197-200
21. Kuske CR, Kirkpatrick BC and Seemüller E (1991) Differentiation of virescence mycoplasmalike organisms using western aster yellows MLO chromosomal DNA probes and restriction fragment length polymorphism analysis. J Gen Microbiol 137:153-159
22. Lee IM and Davis RG (1992) Mycoplasmas which infect insects and plants. In: J Maniloff, RN McElmansey, LR Finch and JB Baseman, eds. Mycoplasmas: Molecular Biology and Pathogenesis, pp 379-390. Washington: American Society for Microbiology

23. Lee IM, Davis RE, Sinclair WA, DeWitt ND and Conti M (1993) Genetic relatedness of mycoplasmalike organisms detected in *Ulmus* spp. in the United States and Italy by means of DNA probes and polymerase chain reactions. Phytopath 83:829-833
24. Lee IM, Hammond RW, Davis RE and Gunderson DE (1993) Universal amplification and analysis of pathogen 16S rDNA for classification and identification of mycoplasmalike organisms. Phytopath 83:834-842
25. McCoy RE (1976) Uptake, translocation and persistance of oxytetracycline in coconut palm. Phytopath 66: 1038-1042
26. McCoy RE, Cauldwell A, Chang CJ, Chen TA, Chiykowski LN, Cousin MT, Dale JL, De Leeuw GTN, Golino DA, Hackett KJ, Marwitz BC, Petzold H, Sinha R, Sugiura M, Whitcombe RF, Yang IL, Zhu BM and Seemüller E (1989) Plant diseases associated with mycoplasmalike organisms. In: RF Whicombe and JG Tully, eds. The Mycoplasmas Vol. 5. Spiroplasma, Acholeplasmas and Mycoplasmas of Plants and Arthropods. San Diego: Academic Press.
27. Nienhaus F, Schuiling M, Gliem G, Schninzer V and Spittel A (1982) Investigations on the etiology of lethal disease of coconut palm in Tanzania. Z Pflkrankh Pflschutz 89:185-193
28. Nutman FJ and Roberts FM (1955) Lethal yellowing the 'unknown disease` of coconut palms of Jamaica. Emp J Exp Agric 23:257-267
29. Plavsic-Banjac B, Hunt P and Maramorosch K (1972) Mycoplasmalike bodies associated with lethal yellowing disease of coconut palm. Phytopath 62:295-299
30. Renard JL, Quillec G and Arnaud F (1975) Une novelle maladie du cocotiers en pepiniére. Oléagineux 30:109-112
31. Rohde W, Kullaya A, Mpunami AA and Becker D (1993) Rapid and sensitive diagnosis of mycoplasmalike organisms associated with lethal disease of coconut palm by a specially primed polymerase chain reaction for the amplification of 16S rDNA. Oléagineux 47: 511-515
32. Romney DH (1983) Brief review of coconut lethal yellowing. Indian Coconut J 13:1-8
33. Schaff D, Lee IM and Davis RE (1992) Sensitive detection and identification of mycoplasma-like organisms in plants by polymerase chain reactions. Biochem Biophys Res Commun 186:1503-1509
34. Seemüller E (1976) Demonstration of mycoplasmalike organisms in the phloem of trees with pear decline or proliferation symptoms by fluorescence microscopy. Phytopath Z 85:368-372
35. Shiomi T and Sugiura M (1984) Grouping of mycoplasmalike organisms transmitted by the leafhopper vector *Macrosteles orientalis* virraste, based on host range. Ann Phytopath Soc Japan 50:149-157
36. Schuiling M, Kaiza DA and Mpunami AA (1992) Lethal disease of coconut palm in Tanzania II. History, distribution and epidemiology. Oléagineux 47:516-521
37. Schuiling M and Mpunami AA (1990) The lethal disease of coconut palm in Tanzania: review of research up to date and preliminary results of resistance trials. In: ML Robert and D Zizumbo, eds. La problemistica del amarillamiento letal de cocotero en México. pp 171-183. Mérida: Centro de Investigacian Cientifica de Yucatan
38. Schuiling M, Mpunami AA, Kaiza DA and Harries HC (1992) Lethal disease of the coconut palm in Tanzania III. Low resistance of imported germplasm. Oléagineux 47:693-697
39. Sinclair WA, Iuli RJ. Dyer AT and Larsen AO (1989) Sampling and histological procedures for diagnosis of ash yellows. Plant Dis 73:432-435
40. Solomon JJ and Govindankutty MP (1983) Association of mycoplasma-like organism with the coconut root (wilt) disease in India. J Plant Dis Protect 90:295-297
41. Turner PD, Jones P and Kenten RH (1978) Coconut stem necrosis a disease of hybrid and Malayan Dwarf coconuts in north Sumatra and peninsular Malaysia. Perak Planters´J pp. 33-46.

Lethal yellowing vector studies. I. Methods of experimental transmission

FORREST W. HOWARD
University of Florida, Fort Lauderdale Research and Education Center, 3205 College Avenue, Fort Lauderdale, Florida 33314, USA

1. Introduction

Since the publication of the first papers on lethal yellowing (LY) transmission by *Myndus crudus* Van Duzee [13, 15], researchers familiar with the LY problem generally consider this species to be a vector, and no evidence has been published to seriously challenge this concept. Based on general knowledge about insect-vectored plant and animal diseases, the possibility that LY is transmitted by more than one vector is highly plausible. However, thus far, there is no evidence for additional vector species.

In my opinion, *M. crudus* (Fig. 1) is probably the only vector of LY, at least in Florida. If there are additional vectors in Florida, they must be rare, localised, or otherwise of minor importance in the epidemiology of this disease. *Myndus crudus* is widely distributed and frequently found on palms in southern Florida. The other auchenorrhynchous insect species associated with palms in this region are distributed more sporadically or locally. In other areas of the Americas affected by LY (*e.g.* México and Jamaica), there are some auchenorrhynchous insect species on palms that are not present in Florida. However, there is presently no evidence to focus on any one of these species as possible vectors of LY.

Knowledge of LY-transmission is still very rudimentary, and further research to elucidate it should be encouraged, especially in regions other than Florida. To encourage such efforts, I have reviewed the methods that we used in successful LY-transmission experiments with *M. crudus* in Florida.

C. Oropeza et al. (eds.), Lethal Yellowing: Research and Practical Aspects, 43–57.

Figure 1. Myndus crudus Van Duzee. Photograph by J.V. DeFilippis, University of Florida.

2. A strategy for searching for a putative vector

Our strategy to identify a vector of LY was to first study the insects associated with palms in areas affected by LY to determine which species appeared most likely to spread the disease, and then test the most likely candidates by transmission experiments. The preliminary field work was important in narrowing down our choices to a minimum of suspected vector species. We wished to avoid testing unlikely candidates since we predicted that the transmission experiments would be particularly arduous. We visualised three particularly difficult obstacles.

Firstly, it was thought that the test palms had to be large. Based on observations of several authors [2, 3, 4, 17], palms less than three years old were rarely susceptible, and the susceptibility of palms increased with age and/or height. Thus, we decided to use the largest test palms that we could practically manage under experimental conditions. It would be necessary to locate palms of several years of age in an area far outside the area affected by LY, negotiate with the owner to obtain them, transport them without damaging or exposing them to disease transmission, plant them in cages without losing any to "transplant shock", and maintain them in healthy condition for a long period. The large cages that would be needed to contain these palms would have to be carefully designed and maintained in perfect condition to confine the insects to be tested, and to exclude others.

Secondly, because of the long incubation period, which was estimated to be 114-450 days [1, 5, 18], we expected a transmission experiment to take at least a year.

Lastly, a successful transmission experiment would require large numbers of test insects. Since nothing was known about vector acquisition of the MLO (*e.g.* the age of the leaves they fed on, the duration of the feeding period, *etc.*), we planned to use wild-caught specimens, a portion of which were presumably infected with LY. Only a portion of these might be expected to transmit the disease agent under our experimental conditions. We only had notions of the conditions conducive to transmission, but surmised that under our experimental conditions, transmission was likely to be inefficient. Thus, to compensate for the inefficiency that would be inherent to this kind of experiment, we would use massive numbers of insects. Collecting and identifying these could be labour-intensive.

3. Focusing on a suspected vector

Since LY first became a threat to coconut production in the Caribbean in the late 1800s, many kinds of organisms have been suspected and, in some cases, investigated as possible vectors. Most of these were species of different orders of insects. Nematodes had been investigated without positive results [17], nevertheless, the idea that they could be the principal vectors persisted through the 1970s. At a meeting of LY researchers in Florida in the 1970s, one invited expert proposed that mites (Acari) had been largely overlooked as possible vectors and should be investigated. A popular idea in Florida, where the disease affected ornamental palms grown in urban areas, was that pigeons, or other kinds of birds common to such environments, spread the disease. This was reminiscent of notions in the Caribbean early in this century that "John-crows" (vultures) played this role. Combining these two ideas, a notion was briefly discussed that mites spread from palm to palm by birds could possibly transmit the disease.

Fortunately, when we were commencing our studies to discover a vector of LY, plant pathologists and other researchers provided an important clue to the mode of transmission: evidence that the causal organism was a mycoplasma-like organism (see Eden-Green, this volume). It was known that plant diseases associated with MLOs were generally transmitted by insects of the leafhopper group, *i.e.* Auchenorrhyncha of the order Hemiptera [6]. *Myndus crudus* was identified as the most common auchenorrhynchous insect on palms in LY-affected areas of Florida [22] and Jamaica [19].

Conditions in Florida offered exceptional opportunities to make certain observations of LY. One advantage was the broad array of palm species present in the LY-affected area. Wherever LY had struck in the Caribbean, it had been known as a disease almost exclusively of coconut palms. This was probably partly because coconut palms were grown in massive monocultures and few additional exotic species of palms were grown. In contrast, in southern Florida, while coconut palm was one of the predominant species grown as an ornamental plant in urban areas, many additional exotic species of palms were distributed in the landscape. The area thus served as a test garden for defining, albeit imperfectly, the host plant range of LY. Over 30 species of palms were found to be susceptible [14, 20, 21]. This raised the question of whether any species of insect visited all of the species of palms susceptible to LY. A survey was conducted and *M. crudus* was found on nearly all of the susceptible palm species, the exceptions being that some of the palms were too rare in Florida to be sampled adequately [10].

Progress was also made due to the fortuitous abundance of an exotic species, *Veitchia merrillii* (Beccari) H. E. Moore (Manila or Christmas palm) in the urban areas of southern Florida. This species is relatively small and fast-growing, reaching a maximum height of approximately 5 metres and with much smaller fronds than the coconut palm. It quickly proved to be one of the more susceptible palm species. Its relatively fast growth and small size at maturity made it a suitable plant for LY-transmission research. For example, in an experiment to test the impact of insecticidal applications on the spread of LY, Manila palms were selected as the test plant because of the ease with which they could be treated (Fig. 2) [9]. The experiment provided evidence that *M. crudus* is the vector because the treatments slightly reduced both the populations of this insect on palms and the spread of LY. Also, because of their small stature, Manila palms were easy to transport, transplant and maintain in transmission cages [11].

During the 1970s and early 1980s, the area affected by LY on the southeast coastal region of Florida was well defined. This situation was ideal for a field study. Population levels of this insect were 40 times higher in the LY-affected zone than in similar areas outside of it [8].

Finally, the diversity of insects associated with palms in Florida is somewhat lower than I generally observed in Caribbean countries, *e.g.* the Dominican

Figure 2. Application of insecticide to a Manila palm, *Veitchia merrillii*, with handheld sprayer Photograph by Gloria Howard.

Republic [12]. This is not unexpected, since southern Florida represents the northern extent of the ranges of coconut and most other palms. In southern Florida, there are only three species of auchenorrhynchous insects found of sufficient frequency on coconut and the other common susceptible palms to imply a consistent association. These are, in addition to *M. crudus*, *Cedusa inflata* Ball (Derbidae) and *Idioderma virescens* Van Duzee (Membracidae). Results of field work gave us reasons to suspect *M. crudus*, but not the other two, as natural vectors of LY.

4. Review of LY-transmission experiments

Three separate transmission experiments were conducted by our team. The first was a preliminary test, an initial probe with *M. crudus* as the suspected vector to seek some indication of whether more conclusive, but more elaborate experiments were warranted. Conducting this experiment was also a way of developing methods for future transmission experiments. The experiment was conducted using a very large cage divided into two sections (Fig. 3). We planted approximately 15 palms of susceptible species in each section. In one section, we introduced *M. crudus* captured from palms in LY-affected areas. During the period June 1977 - September 1978, hundreds to thousands of *M. crudus* were introduced each week, for a total of 20 000 of these insects. The other section served as a control. One coconut palm and two Manila palms contracted LY, and these were in the section where we had introduced *M. crudus*. The diagnoses were confirmed by observing MLO in tissue samples from these palms. The results seemed to indicate that we were on the right track. However, because this was not a replicated experiment, the results were not conclusive and were not published.

Figure 3. The large cage used for the preliminary transmission test with *Myndus crudus*, Fort Lauderdale, Florida, USA.

A replicated experiment [13], which began in September 1978 and continued for 34 months, was designed with five replications, *i.e.* five cages into which we introduced *M. crudus* collected from palms in LY-affected areas, and five control cages which remained free of these insects (Fig. 4). A minimum of one palm in each of five cages where we had introduced *M. crudus* contracted LY (treatment cages). Three of five coconut palms, five of seven Manila palms and two of three *Pritchardia thurstonii* F. Müller et Drude contracted LY in the treatment cages. By comparison, no palms in the control cages were affected by LY. All diagnoses of LY were confirmed by observation of MLOs in tissue samples examined by electron microscopy (EM). No LY symptoms developed in the palms in the control cages. At the end of the experiment, the latter were sampled for EM and found to be free of MLOs.

Figure 4. One of 10 cages used in the replicated transmission experiment with *Myndus crudus*, Fort Lauderdale, Florida, USA.

A third experiment was performed to determine whether LY could be transmitted to young palms of susceptible species [15]. A total of 98 palms of 9 species were divided among two cages into which *M. crudus* was introduced. A result of this experiment was that LY was transmitted under experimental conditions to small potted palms less than two years old, including two coconut palms, two *Pritchardia pacifica* Seeman and Wendland and a Chinese windmill palm (*Trachycarpus fortunei* (W.J. Hooker) J. Wendland). Additionally, LY was transmitted to three palms of three to four years, including one *Phoenix canariensis* Hort. ex Chab. and two *Pritchardia thurstoni*. The *T. fortunei*, the youngest palm to which LY was transmitted in this experiment, was 14 months old and 30 cm tall (measured from the base of the stem to the tallest leaf) when LY symptoms were first visible.

5. Methods used in LY-transmission experiments

Those three rather crude experiments yielded the first convincing results that a particular species of insect transmits LY. Future transmission experiments, which will be essential for further elucidating the nature of this disease, will be more sophisticated as new knowledge and techniques become available. However, some of our methods or modifications of them may be applicable to future research on LY-transmission.

Undoubtedly, one of the most important factors that contributed to positive results in our transmission experiments was the large number of test insects involved. This is the one point that has most often been misunderstood by colleagues. The argument is often presented that, because "it took up to 20 000 *M. crudus* to achieve transmission of LY" that this could not be the vector in nature, or it is a highly inefficient vector and there must be other, more efficient vectors. It is therefore important that I emphasise that we know only the total number of insects that we captured and transferred to cages, but our experiments provided no information whatsoever on the number of insects involved in transmission or the percentage of the population that was infected.

It can be assumed that the percentage of infection in a vector species differs in different populations and in the same population may fluctuate, depending on a host of environmental factors that are also highly variable. At any given time within the infected portion of the population, an unknown percentage of individuals would have become infected too recently to transmit the agent, or conversely would be past that stage. Within the portion of the population that is capable of transmitting the infectious agent, chance would dictate which portion of them fed on susceptible palms under the conditions necessary for transmission to take place. Thus, testing a suspected vector *via* a transmission experiment is similar to playing a game of chance without knowing the chances of winning.

Gamblers may increase their chances of winning a lottery by buying more tickets. Similarly, part of our strategy was to increase the chances of a transmission (assuming that we were correct in our choice of species) by employing a maximum number of insects. Actual numbers of *M. crudus* transferred in the three experiments can be prorated to show approximate numbers per cage or per palm per month:

First experiment: 20 000 *M. crudus* (accumulative total) in one cage section for 15 months, *i.e.* an average of about 1233 (89 per palm) per month.

Second experiment: approximately 30 000 *M. crudus* per cage (accumulative total), *i.e.* about 850 per cage (*i.e.* approximately 280 per palm) per month for 34 months.

Third experiment: cage #1, containing 84 small palms: approximately 1 900 *M. crudus* per cage (approximately 23 per palm) per month for 18 months (April 1981-October 1982). Cage #2, containing 14 small palms: approximately 500 *M. crudus* per cage (approximately 36 per palm) per month for 11 months (November 1981-September 1982).

More important than the number of test insects that were transferred to the cages is the number that actually fed on the palms. It was not possible to determine this number for any experiment, but it appeared to be a relatively low number compared with the numbers that could be collected from palms *in situ*. Whenever we introduced *M. crudus* into a cage with palms, most of them usually flew to the sides of the cage and remained there for long periods. Often when we inspected the palms later, fewer *M. crudus* were resting or feeding on them than we would have expected, based on the number that we had introduced. Many hours of collecting these insects taught us that the numbers of *M. crudus* that visit palms in nature are highly variable. We have collected hundreds from a single palm and we have examined many palms on which we could find no specimen of this insect. Presumably the frequency that they move from palm to palm, and the frequency that these insects visit a palm during a month varies greatly depend on many factors. Thus, a relatively low number of *M. crudus* apparently occupied palms in our cages, as often was the case under natural conditions. However, under natural conditions, much higher numbers of *M. crudus* are often seen on palms *in situ* than seen in the cages.

Our general procedure was to continue introducing test insects as long as symptomless palms were present in a cage, or until several palms had contracted LY. The transmissions would have occurred at some unknown point in time, probably several months prior to the appearance of symptoms. Assuming that a single insect transmits enough inoculum to initiate an infection, any transmissions after that would have been superfluous. This would represent still another portion of the test insects that could not have played a role in transmission.

It is not known whether LY transmission is influenced by seasonal factors in Florida. For this reason, we collected during every month of the year. This procedure contributed to a high accumulative total of test insects. But if LY transmission is highly seasonal, the test insects transferred during the rest of the year were superfluous.

If it was known that vectors of LY were most likely to become infected when feeding on leaves in a particular stage of symptom development, *e.g.* yellow leaves of palms in early stages of symptom development, it would have been reasonable to collect all of the test insects from such sites. However, such information was not known. In the absence this knowledge, pre-symptomatic infected palms seemed as likely an infection source as symptomatic palms. However, we had no means to detect the former. Thus, to increase the chances that some of our test insects would be infected, we collected them from symptomatic palms and palms adjacent to these, and from a large range of sites on palms.

In summary, our experiments involved transfer of large numbers (*i.e.* accumulative totals) of test insects from the field to experiment cages, but these numbers do not seem excessive when prorated by month, cage, and palm and compared to field situations. Furthermore, the total insects involved in experiments should not be assumed to represent the number needed to accomplish transmission. We can assume that the transmission results would have been the same in any particular cage whether 20 000 *M. crudus* or 0.005 % of this number (*i.e.*, a single *M. crudus*) had transmitted the disease agent. Large numbers of test insects from diverse situations were used to increase the probability of a transmission. A large portion of the test insects can be assumed to have been superfluous to the experiment, partly because a large portion apparently underwent dispersal behaviour and either did not feed on palms, or probably did not feed normally. A proportion of them would have been introduced into cages after transmission had already taken place. If infection rates peaked seasonally, the portion introduced during much of the year may have had a low infection rate. In short, the main reason so many insects were used was because, aside from the evidence that seemed to implicate *M. crudus* as the vector, nothing was known about transmitting this disease. As more is learned, it should be possible to accomplish transmission with fewer test insects.

Aspirators are often used in entomological work, including studies of insect-vectored plant diseases, because they can be used to collect large numbers of insects quickly. However, the effect of this technique is that the insects, resting on the plant surfaces and in many cases with their stylets inserted in plant tissue, are suddenly wrenched from that surface and hurled through a long tube and against the side of the collecting jar. Since it was critical to successful transmission that they be capable of normal feeding behaviour, we designed a method that would induce the insects to leave the plant surface passively. This "passive collecting device" was a stoppered bottle with a tube inserted through the stopper (Fig. 5) The mouth of the tube was placed over or just anterior to an insect identified as *M. crudus* resting or feeding on a palm leaf surface. The response of the insect was to jump or fly, sometimes after a brief pause. The tube directed their flight into the collecting bottle and served as a funnel to trap them there. We kept a piece of leaflet or crumpled tissue paper in the bottle to give them landing and resting sites. Because we had found that *M. crudus* does not survive well in vials or bottles exposed to direct sunlight, the bottle was partially wrapped in duct tape, leaving small windows for viewing the insects.

Collecting with this device is difficult at first, but our employees became quite skilful, commonly collecting 1 000 or more *M. crudus* in a period of a few hours. Usually they took several bottles when collecting, or made multiple releases into cages so as not to accumulate excessive numbers of insects in one bottle.

An additional advantage of this device over aspirators is that because the activity of collecting each insect is more deliberate, trained collectors are less likely to collect the wrong insect. To further insure that only *M. crudus* was transferred to cages, when the collecting was done by a new employee, each catch was examined before it was released in a cage. After an employee had collected for a few months under close supervision, the catches were only checked at intervals.

It is not known whether use of the passive collecting device was important to transmission results. This method remains an option which insures care in identification and handling of the test insects at the expense of the higher numbers that can be collected by aspirators.

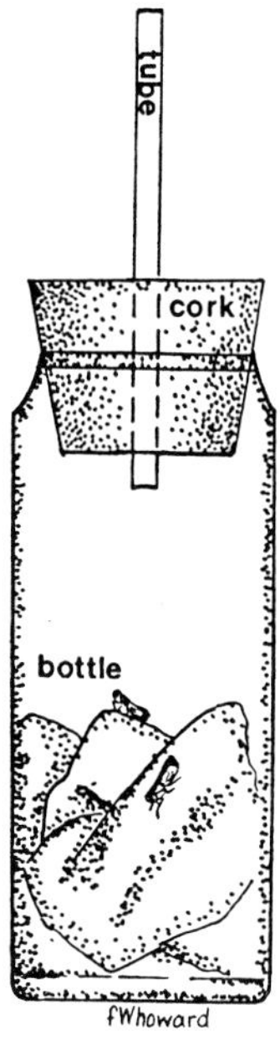

Figure 5. A simple device for passive collecting of planthoppers

Several factors that we initially thought might be important in transmission experiments of a suspected LY vector could be discounted on the basis of our results. Initially, we thought that it would be necessary to use large palms, based on field observations of several authors [2, 3, 4, 17]. We learned from our third experiment that transmission results can be obtained using small potted palms. This will greatly facilitate future transmission experiments.

In spite of early success with Manila palm, the results of these experiments eventually suggested that any susceptible palm species may be suitable as a test plant. We obtained positive transmission results with young palms of several diverse genera (Table 1).

Table 1. Palm genera suitable for LY-transmission studies

Subfamily	Tribe	Subtribe	Genus
Coryphoideae	Corypheae	Livistoninae	*Pritchardia*
		Thrinacinae	*Trachycarpus*
	Phoeniceae		*Phoenix*
Arecoideae	Areceae	Ptychospermatinae	*Veitchia*
	Cocoeae	Butiinae	*Cocos*

In the interest of simulating field conditions, St. Augustine grass, *Stenotaphrum secundatum*, a preferred breeding host of *M. crudus* [7], was planted and maintained in the cages used in the first two experiments. To insure that the grass was not contaminated by insects, we requested the US Department of Agriculture to treat the sod pieces in a fumigation chamber. Unfortunately, this treatment killed the grass. Thus, before planting the grass in cages, we washed and carefully examined each stolon with a magnifying lens. However, the third experiment, in which transmission took place in cages devoid of grasses, demonstrated that these extraordinary efforts were unnecessary.

Unpublished casual observations of several researchers have suggested that palms under shade are less likely to contract LY than palms in full sun. To take this possibility into account, we designed our cages so as to permit maximum penetration of sunlight, but minimum access to insects. Clear polyethylene was used for the roof and the bottom half of each side. This material was not used for the entire cage because it would have resulted in excessive interior temperatures. To prevent this, the upper half of each side was of fibreglass screen. We chose an opening of 0.9 mm, which we felt allowed ample entry of light, but did not permit passage of *M. crudus* or larger insects. It is not know whether providing for maximum penetration of sunlight was a significant factor in transmission by *M. crudus*. Since this seems plausible, I recommend that it be considered in future transmission experiments.

As long as transmission experiments are conducted using field cages, great care must be exercised to prevent insects other than test insects from entering the cages. In addition to limiting access by choice of cage roof and size materials as discussed, burrowing insects and some ants were excluded by extending the cage sides 10 cm into the soil. Rather than install, *e.g.* wooden framed doors which often warp and thus provide access for insects, we designed a door more apt to provide a "tight seal". It consisted of a semicircular slit in the polyethylene side that was opened and closed by a zipper (Fig. 4). A brush was available in the cage area with which workers brushed themselves off prior to entering a cage to avoid introducing contaminant insects. The cages were inspected daily. Small tears or holes developed only occasionally and were immediately repaired. The palms and other aspects of the cage interiors were frequently examined. Contamination by some very small arthropods, notably scale insects (Diaspididae) and mites (Acari), was practically unavoidable. However, these could be discounted as vectors because of their known biology and, in the replicated (second) experiment, they were distributed in both treatment and controls. It is encouraging that in third experiment there were fewer contaminants. This was probably because smaller palms were used, and possibly because of the absence of grass in the cages.

The use of smaller palms and the availability of more advanced diagnostic techniques (see Harrison *et al.*, this volume and Jones *et al.*, this volume) will greatly facilitate transmission experiments in the future. Hopefully, future researchers will also benefit from our experiences in conducting LY-transmission research.

Acknowledgements

I wish to record that during the 1970s and early 1980s, those of us who were investigating LY in Florida were in close touch with a first-rate research group in Jamaica sponsored by the British Overseas Development Administration. The small advances that we made in Florida in understanding the nature of LY-transmission were greatly facilitated by the amiable co-operation and encouragement of our colleagues in Jamaica.

References

1. Britt LL (1981) A study in spatial diffusion: lethal yellowing of *Cocos nucifera*. MA thesis, University of Miami
2. Bruner SC and Boucle L (1943) La enfermedad conocida por "pudricion del cogollo" del coctero en Cuba. Rev Agric 26:122-141
3. Carter W (1966) Susceptibility of coconut palm to lethal yellowing disease. Nature 212:320
4. Carter W and Suah JRR (1964) Studies on the spread of lethal yellowing disease of the coconut palm. FAO Plant Protect Bull 12:73-78

5. Dabek AJ (1975) The incubation period, rate of transmission and effect on growth of coconut lethal yellowing disease in Jamaica. Phytopath Z 84:1-9
6. D'Arcy CJ and Nault LR (1982) Insect transmission of plant viruses and mycoplasmalike and rickettsialike organisms. Plant Dis 66:99-104
7. Eden-Green S J (1978) Rearing and transmission techniques for *Haplaxius* sp., a suspected vector of lethal yellowing disease of coconut palms in Jamaica. Ann Appl Biol 89:173-176
8. Howard FW (1980) Population densities of *Myndus crudus* Van Duzee (Homoptera: Cixiidae) in relation to coconut lethal yellowing distribution in Florida. Principes 24:174-178
9. Howard FW and McCoy RE (1980) Reduction in spread of mycoplasma-like organism associated lethal decline of the palm *Veitchia merrillii* by use of insecticides. J Econ Entomol 73:268-270
10. Howard FW and Mead FM (1980) A survey of Auchenorrhyncha (Insecta: Homoptera) associated with palms in southern Florida. Trop Agric 57:145-153
11. Howard FW and Thomas DL (1980) Transmission of palm lethal decline to *Veitchia merrillii* by a planthopper, *Myndus crudus*. J Econ Entomol 73:715-717
12. Howard FW, Kramer JP and Feliz-Peralta M (1981) Homópteros Auchenorrhyncha asociados a palmeras en un área de la República Dominicana afectada por el amarillamiento letal del cocotero. Folia Entom Mexicana 47:37-50
13 Howard FW, Norris RC and Thomas DL (1983) Evidence of transmission of palm lethal yellowing agent by a planthopper, *Myndus crudus* (Homoptera: Cixiidae). Trop Agric 60:168-171
14. Howard FW, Thomas DL, Donselman HM and Collns ME (1979) Susceptibilidad de algunas especies de palmeras a enfermedades asociadas a organismos micoplasmöides en Florida. FAO Bol Fitosan 27:109-117
15. Howard FW Williams DS and Norris RC (1984) Insect transmission of lethal yellowing to young palms. Int J Entomol 26: 331-338
16. Latta RK (1966) Attempts to relate nematodes to lethal yellowing disease of coconut palms in Jamaica. Trop Agric 43:59-68
17. Nutman FJ and Roberts PM (1955) Lethal yellowing: the "unknown disease" of coconut palms in Jamaica. Emp J Exp Agric 23:257-267
18. Romney DH (1972) Past studies on and present status of lethal yellowing disease of coconuts. PANS 18:386-395
29. Schuiling M (1976) A survey of insect populations on *Cocos nucifera*. (Abstr.) Principes 20:67
20. Thomas DL (1979) Mycoplasmalike bodies associated with lethal declines of palms in Florida. Phytopath 69:928-934
21. Thomas DL and Norris RC (1981) The use of electron microscopy for lethal yellowing diagnosis. Proc Florida State Hort Soc 93:196-199
22. Woodiel NL (1976) Insects associated with coconut palm in south Florida. (Abstr.) Principes 20:66-67

Lethal yellowing vector studies. II. Status of *Myndus crudus* host plant studies

FORREST W. HOWARD
University of Florida, Fort Lauderdale Research and Education Center, 3205 College Avenue, Fort Lauderdale FL 33314, USA

1. Introduction

Myndus crudus Van Duzee appears to be the major, if not sole, vector of LY in Florida (Howard, this volume). Recently, we have been attempting to manage populations of this insect by management of the ground cover under palms.

In most Hemiptera and other exopterygote orders, the immature stages occupy the same hosts as the adults. The hemipterous family Cixiidae, to which *M. crudus* belongs, is unusual in that the immature stages generally develop on herbaceous plants, often in the root zone, and the adults occupy arborescent hosts. *M. crudus* appears to be associated mostly, or perhaps exclusively, with monocotyledonous plants. The adult hosts appear to be mainly palms [6]. They are also found on *Pandanus* spp. (Pandanaceae) (Howard, unpublished), an arborescent monocotyledonous plant that is susceptible to Caribbean-type LY (Harrison, this volume), while those of the immature stages ("breeding hosts") are thought to be almost exclusively grasses of many species [9, 11, 12]. A dicotyledonous plant, *Verbena scabra* Vahl (Verbenaceae), was included in a host plant list of the species [11], but its status as a possible host requires further confirmation and clarification. In an area of southern Florida where *M. crudus* was found in the root zone of St. Augustine grass (*Stenotaphrum secundatum* (Walt.) Kuntze), 100-200 specimens of each of six of the most common herbaceous dicotyledonous weeds were examined without finding *M. crudus* immature stages [8]. This preliminary data requires confirmation.

St. Augustine grass has long been known to be a preferred breeding host of *M. crudus* [1, 10]. As this is the most common turf-grass in the urban areas of

C. Oropeza et al. (eds.), Lethal Yellowing: Research and Practical Aspects, 59–63.

Southeastern Florida, it has been suspected to be a major factor in the spread of LY in this area [2]. Guinea grass (*Panicum maximum* Jacquin) is also an effective breeding host for *M. crudus* [4], and is probably the most common grass in Jamaican coconut plantations (Been, pers. comm.).

Casual observations in Florida have suggested that LY may spread relatively slowly in areas devoid of vegetation, such as large paved parking lots and extensive sandy ocean beaches bordered by streets and buildings, and relatively rapidly in grassy areas such as golf courses. It is thus hypothesised that the spread of LY could be reduced in plantations by managing the herbaceous ground cover to reduce or eliminate breeding hosts of *M. crudus*, and conserve species that serve as poor breeding hosts. This technique could be used to reduce disease pressure against LY resistant palm varieties and hybrids.

The components of our research plan to identify poor breeding hosts of *M. crudus* are :

(a) selection of herbaceous species, including grasses and dicotyledonous plants, that are desirable as forage, turf-grass or ornamental ground covers,

(b) tests using containerised specimens of the selected species,

(c) small field plot tests of plants selected from the tests with containerised plants, and

(d) tests with large plots under farm conditions to determine the actual effect of the method on reducing populations of *M. crudus*.

2. Tests with plants in containers

The objective of using containerised plants is to maximise the number of species tested with a minimum of space and resources. One method was to grow the test plants in small (*e.g.* 7 cm diameter) containers and maintain these in a cage with wild-caught *M. crudus* adults. The plants in these containers were more readily colonised by *M. crudus* when the insects were concentrated in cages than when the containerised plants were exposed to wild populations in the field. However, collecting the insects is labour-intensive. Recently, we have found that grasses grown in 20 cm x 30 cm greenhouse flats exposed outdoors to wild populations are readily colonised.

In container tests, the following grass species were favourable breeding hosts for *M. crudus*: (a) *Cenchrus cilaris* L., (b) *Cynodon nlemfluensis* Vander. *nlemfluensis* cv. 'Puerto Rico star', (c) *Digitaria eriantha* Steud., (d) *Panicum purpurescens* Radii, (e) *Panicum maximum* Jacquin, (f) *Paspalum notatum* Flugge, (g) *Stenotaphrum secundatum* (Walt.) Kuntze cvs and genetic lines: 'Floratam', FX 33, FX 313, FX 216, FX 332 and (h) *Zoisia* sp..

Among these species, *C. nlemfluensis*, *P. notatum* and *S. secundatum* supported the greatest numbers of *M. crudus* immature stages.

The following species were less favourable breeding hosts for *M. crudus*, compared with *S. secundatum* 'Floratam' as a standard: (a) *Axonopis* sp., (b) *Brachiaria brizantha* (Hoschst. ex A. Rich.), (c) *B. humidicola* (Rendle) Schweick, (d) *B. dictyoneura* (Fig. y DeNot) Stapf., (c) *Chloris gayana* Kunthe, (d) *Hemarthria altissima* (Poir) Stapf y C.D. Hubb cvs and genetic lines: 'Floralta', 'Bigalta', PI No. 364871, (e) *Pennisetum purpuretum* Schumacher and (f) *Saccharum officinarum* L. sugarcane selection no. 701527.

3. Small field plot tests

In studies thus far completed, grasses selected as poor breeding hosts in tests with containerised plants were established in replicated plots of 2 m^2. After allowing for exposure to wild populations of *M. crudus* adults, and completion of a life-cycle, modified Malaise-type traps were installed over the plots to collect adults emerging from the grass.

Three relatively poor breeding hosts were selected from the preliminary tests: *B. brizantha*, *C. gayana*, two of which are widely used as tropical forage grasses, and *H. altissima* No. 364871, which is a promising grass for such use.

The results approximated those of the container tests, *i.e.* the three grasses were unfavourable for *M. crudus* development compared to St. Augustine grass. We interpreted these results as indicative that container tests were reliable indicators of the *M. crudus* breeding host potential of grass species. These results were encouraging, but more recently we have observed large numbers of *M. crudus* immature stages associated with a selection of *B. brizantha* grown in a greenhouse flat and exposed to wild populations of the insect. This situation is being investigated.

In another study, it was found that populations of *M. crudus* varied with the length of St. Augustine grass; keeping grass mowed short reduced populations [7].

At present, we are testing leguminous ground covers as possible breeding hosts of *M. crudus*. Leguminous ground covers are often maintained in coconut plantations, and some species appear to have high potential as ornamental ground covers. In another study, we are determining whether the presence of a layer of organic debris on the ground is conducive to development of *M. crudus* immature stages in the grasses, as suggested by earlier work of Eden-Green [1].

Further research is required to elucidate the host plant relationships of *M. crudus* and to use this knowledge in managing this insect.

Acknowledgements

The personal communication of Basil Been of the Coconut Industry Board, Jamaica, is gratefully acknowledged.

References

1. Eden-Green SJ (1978) Rearing and transmission techniques for *Haplaxius* sp., a suspected vector of lethal yellowing disease of coconut palms in Jamaica. Ann Appl Biol 89:173-176
2. Howard FW (1980) Population densities of *Myndus crudus* Van Duzee (Hemiptera: Cixiidae) in relation to coconut lethal yellowing distribution in Florida. Principes 24:174-178
3. Howard FW (1989) Evaluation of six species of grasses as breeding hosts of *Myndus crudus*, a vector of lethal yellowing of palms. Proc Caribbean Food Crops Soc 25:433-438
4. Howard FW (1990) Evaluation of grasses for cultural control of *Myndus crudus*, a vector of lethal yellowing of palms. Entom Exp Appl 56:131-137
5. Howard FW (1990) Evaluation of *Brachiaria* and St. Augustine grasses as breeding hosts of *Myndus crudus* Van Duzee (Hemiptera: Cixiidae), a vector of the lethal yellowing disease of palms. Betr Trop Landwirt Vet Med 28:425-430
6. Howard FW and Mead FM (1980) A survey of Auchenorrhyncha (Insecta: Hemiptera) associated with palms in southern Florida. Trop Agric 57:145-153
7. Howard FW and Villanueva Barradas J (1994) The abundance of the American palm cixiid in relation to grass height and location of the nymphs in relation to soil surface. Z Angewandte Entom 118:239-244
8. Howard FW, Tilmant JR, Burch DG and Hall D (1984) A field study to determine potential host plants of lethal yellowing agent. Report FL-84-1, University of Florida, Fort Lauderdale Res Educ Cent

9. Piña Quijano PE (1992) Estudio poblacional de *Myndus crudus* Van Duzee (Hemiptera: Cixiidae) vector del amarillamiento letal del cocotero en el estado de Yucatán. Lic Biol thesis, Universidad Autonoma de Yucatán, Mérida
10. Reinert JA (1980) Phenology and density of *Haplaxius crudus* (Homoptera: Cixiidae) on three southern turfgrasses. Environ Entom 9:13-15
11. Tsai JH and Kirsch OH (1978) Bionomics of *Haplaxius crudus* (Hemiptera: Cixiidae). Environ Entom 7:305-308
12. Zenner de Polania I and Lopez A (1977) Apuntes sobre la biologia y habitos del *Haplaxius pallidus*, transmisor de la "marchitez sorpresiva" en palma africana. Rev Colombiana Entom 3:49-62

Physiology and biochemistry of lethal yellowing-affected *Cocos nucifera* L. palms

C. OROPEZA, L. ALPIZAR, I. ISLAS-FLORES, A. ESCAMILLA, and J. SANTAMARIA
Centro de Investigatción Científica de Yucatán, Apartado Postal 87, Cordemex, 97310 Mérida Yuc., México

1. Introduction

Lethal yellowing (LY) is a disease which is associated with phloem-inhabiting mycoplasmalike organisms (MLO) [3, 15, 30]. According to Zimmermann [38], if MLOs can only be transported *via* the sieve tubes, movement of MLOs is expected to be from mature leaves to phloem sinks such as roots, flowers, fruits and the shoot apex. Therefore, it is unlikely that many MLOs are present in mature leaves. Evidence of such a distribution of LY-MLOs in coconut (*Cocos nucifera* L.) palms has been obtained using molecular techniques (Escamilla *et al.*, this volume, Harrison *et al.*, this volume). However, leaf senescence, perhaps the most conspicuous symptom in yellows-type diseases, occurs in all leaves, including the mature leaves. Therefore, leaf damage, or the changes therein, probably takes place as an indirect effect of the MLO action. This paper updates a previous review [29] on the physiological and biochemical changes in LY-affected coconut palms and discusses how they may arise by MLO action. An understanding of the physiology and biochemistry of the disease will contribute to a better understanding of the mode of action of the LY-causing MLO, and this in turn, may lead to the development of more effective control measures.

C. Oropeza et al. (eds.), Lethal Yellowing: Research and Practical Aspects, 65–77.

2. Physiological and biochemical changes in shoots

2.1 Disease rating

Different authors studying LY have used different disease rating systems [10, 25]. To discuss data from different authors, two different rating systems are used. These systems, and how they are equated, are presented in Table 1.

Table 1. Disease rating systems for coconut lethal yellowing

Rating systems		Symptoms
Eden-Green [10]	McCoy *et al.* [25]	
Symptomless	0	Healthy or incubating
Early	1	Nutdrop only
	2	One necrotic inflorescence
	3	Two or more necrotic inflorescences
Intermediate	4	Yellowing in lower leaves only
Advanced	5	Yellowing in lower and middle leaves
	6	All leaves yellowed, spear leaf good
Death	7	Spear leaf dead, some green leaves left
	8	Spear leaf dead, all leaves yellowed
	9	Crownless trunk (telephone pole)

2.2 Stomatal behaviour

A symptom consistently found in plants of a number of species affected by MLO-associated yellows, is a reduction in noon stomatal conductance [22], starting at the same time or before the appearance of visual symptoms. This abnormal stomatal closure is complete in lethal yellows diseases, whereas it is partial in nonlethal yellows diseases, suggesting a possible relationship in the mode of action of yellows-MLO and stomatal closure [22].

In coconut palms affected with LY, lower noon stomatal conductance [14, 4] and greater stomatal closure compared with unaffected palms (Table 2), have also been reported. Recent studies carried out in our laboratory [20] have demonstrated the extent of the change in stomatal behaviour. Stomatal conductance normally fluctuates diurnally, however LY decreases conductance in such a way that the

Table 2. Stomata open in median leaves of LY-affected palms at noon [14].

Disease rating	% ± SD
Healthy	82 ± 13
Early	56 ± 06
Advanced	02 ± 02

diurnal fluctuation of stomatal conductance is suppressed (Fig. 1A). This decrease in stomatal conductance starts simultaneously with the onset of symptom appearance (Fig. 1B) and is totally suppressed by stage 3. These results were derived from analysis of middle crown leaves. When other leaves in the crown were studied, their stomatal conductance was also suppressed [20]. Therefore, a

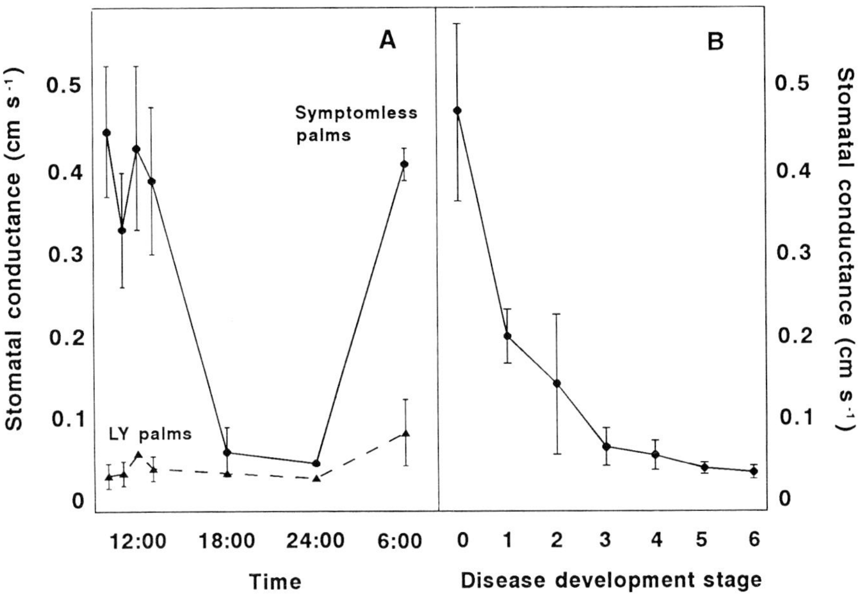

Figure 1.- Diurnal fluctuation of stomatal conductance in symptomless coconut palms and in palms at stage 4 of lethal yellowing development (A). Noon stomatal conductance in coconut palms with different stages of lethal yellowing development (B). Measurements were performed on leaves from the middle of the crown. Bars denote standard deviations. Redrawn from León *et al.* [20].

permanent stomatal closure (even during the light hours) occurs early in the development of LY visual symptoms and it is spread from the base to the top of the crown. Thus, the gas exchange capacity of the crown must be greatly impaired from the onset symptom appearance, if not before. This, in turn, probably affects other functions such as water transport and photosynthesis. Furthermore, Thimann and Satler [35, 36] proposed that stomatal closure triggers leaf senescence, the most conscpicuous symptom of LY (see below).

2.3 Transport

2.3.1 Phloem transport

The presence of MLO in sieve tubes of LY-affected palms has been associated with phloem malfunctions [3]. Eden-Green and Waters [12] found that the inflorescence exudation rate of phloem sap (as toddy collection) decreased in LY-affected palms in early stages of the disease (Fig. 2A). They also reported that

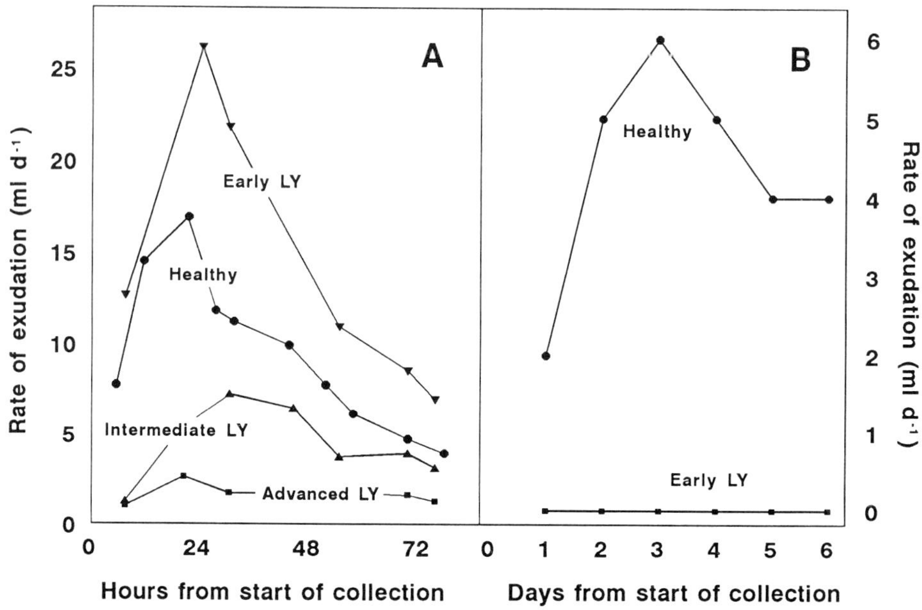

Figure 2. Phloem exudation rates from trunk sections (A) and inflorescences (B) of healthy and lethal yellowing-affected coconut palms. Redrawn from Eden-Green and Waters [12].

the phloem exudation rate of trunk sections increased in palms during early stages, but it was greatly reduced in severely affected palms (Fig.2B). These results suggest that phloem function is altered in LY-affected palms, but does not occur simultaneously in different plant parts. Kollar *et al.* [19] proposed a model for the infection of apple (*Malus domestica*) by MLO that may explain such changes in the rate of exudation. When the MLO population is low, there is impairment of the sieve tube sealing mechanism. As the MLO titre increases, the tubes become thus obstructed reducing phoem flux. This obstruction may be the result of necrosis, callose deposition, or even the MLO themselves occluding the sieve pores. In the phloem of coconut inflorescences, callose deposition at early stages of the disease has been reported [27]. There is also electron [3] and flourescence [4] microscopy evidence to suggest that the MLO might be blocking the sieve tubes. However, an actual blockage in the sieve tubes by the MLO in LY-infected *C. nucifera* palms remains to be demonstrated.

2.3.2 Xylem transport

Presently, there is no evidence of the presence of MLO in the tracheids of LY-affected palms. Neither is there any evidence of any physical obstruction of the tracheidal elements, except in severly diseased palms. There is evidence, however, that water transport through xylem is reduced as a result of LY [5, 6, 14]. Eskafi *et al.* [14] showed that xylem transport was reduced by 75% in LY-diseased palms when ^{32}P was applied at the base of the petiole, and it was reduced by 100% when applied through either stem or roots (Fig. 3). According to the authors, the limiting factor for reduced xylem transport is the closure of stomata. However, root damage has also been reported (see below) and may contribute to reduced xylem transport.

It is important to stress that although evidence discussed above supports the fact that xylem transport is severely affected in LY diseased palms, the data were obtained from palms in the intermediate and advanced stages of the disease. There is no direct data available for early diseased palms. However, McCoy [24] studied the translocation of oxytetracycline in coconut palms and concluded that this antibiotic was distributed primarily by the xylem. McCoy [23] also showed that the beneficial effect of tetracycline antibiotics can be obtained if LY-diseased palms are injected prior to systemic foliar yellowing. These reports suggest that xylem transport is not severely affected in early stages (1 to 3) of the disease.

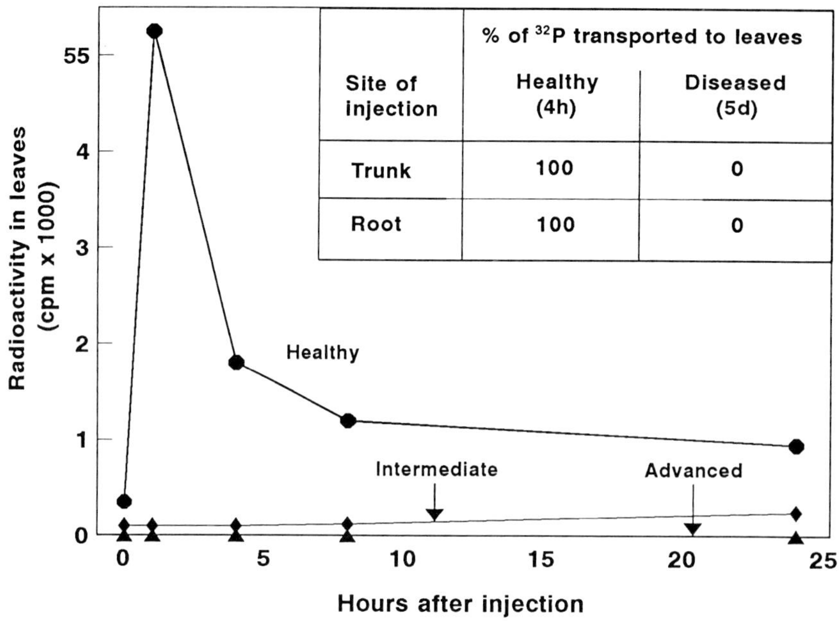

Figure 3. Recovery of ^{32}P from leaves of healthy and lethal yellowing-diseased coconut palms after application at the base of the petiole (graph) and through stem or roots (insert). Redrawn from Eskafi *et al.* [14]

2.4 Water relations

Studies of leaf water relations in LY-affected palms showed no changes before the onset of senescence (stage 5), apart from a small increase in water potential at stage 4 [20]. At stage 5, turgor, osmotic and water potentials decreased [20]. This increase in water potential is consistent with an increase in leaf xylem pressure previously reported for symptomatic palms [26].

2.5 Nutrients

A decreased supply of nutrients should be expected if water transport is impaired. Reduced mineral nutrient supply is hinted at by the ^{32}P transport experiments of Eskafi *et al.* [14]. In addition, Dabek and Hunt [7] reported the effect of exogenous application of various nutrients including Cu and Zn on detatched pinnae from LY-diseased fronds of *C. nucifera* showing yellowing. Both nutreints promoted re-greening, suggesting that the pinnae had a micronutrient deficiency. However, both pieces of evidence are indirect. Direct measurements of the content of mineral nutrients (Cu, Zn, K, N, Mg, Ca and Fe) in the trunk phloem failed to show differences between healthy and diseased palms [32]. Therefore, conclusive evidence of a mineral nutrient deficiency as a result of LY is still lacking.

With respect to organic nutrients, Stemmner *et al.* [34] found no difference in the amino acid composition of the phloem sap exuded from trunks of healthy and diseased coconut. In contrast, Sánchez *et al.* [32], found that the leaf total amino acid and arginine contents in LY-affected coconut palms, was three- and five-fold higher, respectively, than in symptomless palms. These increases may result from protein degradation since a simultanous three-fold protein content decrease accompanies this occurrence [20]. These divergent results may be related to the plant parts studied by each group. Changes in the contents of organic nutrients have also been found in roots (see below).

2.6 Senescence

Changes commonly associated with leaf senescence are decreases in photosynthetic capacity, pigment content and protein content. These parameters were found to decrease in diseased palms [20]. The photosynthetic rate (Fig. 4A) and protein content (Fig. 4B) did not significantly change during early stages of the disease, but dropped sharply in later stages. The carotenoid content decreased steadily contrasting with a sharper decrease in the chlorophyll content, particularly during later stages of the disease. No qualitative changes in pigments were observed (Fig. 4C). A decrease in photosynthetic capacity occurred long after stomatal conductance had been suppressed, suggesting that LY-induced stomatal closure was not limiting CO_2 availability and therefore other factors, such as changes in photosythetic pigments and activity of the enzymes of the carbon reduction cycle, might be involved in the decrease of photosynthetic capacity. The photosynthetic pigments and protein contents decreased simultaneously with photosynthetic rate. The protein loss is likely to be affecting ribulose biphosphate carboxylase (Rubisco) which accounts for 50% or more of the soluble leaf protein in several species [13]. A sharp decrease in an abundant polypeptide (Mr = 55 000) in leaves of diseased palms has been observed (Villanueva, pers. comm.). Taking together, these data suggest that the decline of the photosynthetic rate in LY-affected palms is the result of decreasing contents of pigments and specific proteins rather than a reduced abvailability of CO_2 due to permanent stomatal closure.

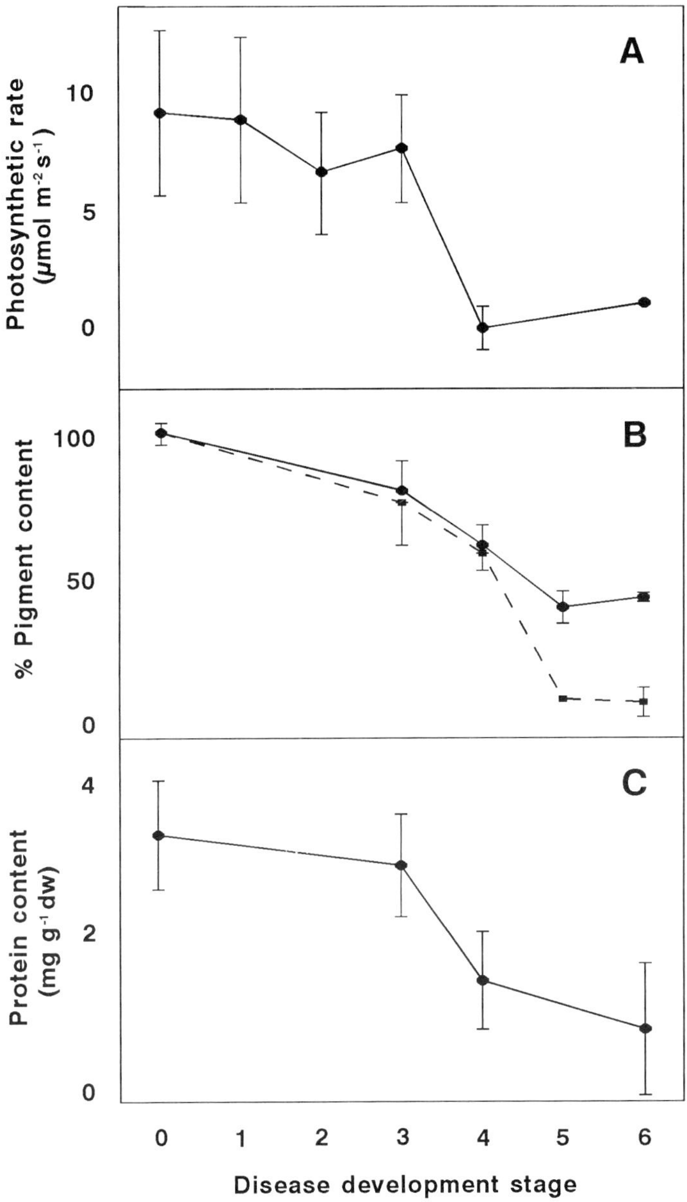

Figure 4.- Photosynthetic rates (A), chlorophyll and carotenoid contents (B), and protein content (C) in leaves of coconut palms at different stages of lethal yellowing development. Measurements were carried out on leaves from the middle of the crown. Bars denote standard deviations. Redrawn from León *et al.* [20].

2.7 *Growth regulators*

Reports discussed above have shown a series of physiological and biochemical alterations in palms affected by LY. Most of these alterations are described in mature leaves. However, since MLO are rarely found in mature leaf tissues, the MLO must be causing indirect effects. This may occur by translocation of metabolites produced by the pathogen or by the induction of the host plant to produce them [38]. Plant growth regulators (PGR) may be such metabolites. Matteoni and Sinclair [22] suggested that a hormonal imbalance could be involved in the MLO action in yellows diseases.

LY symptoms, such as nutfall (fruit abscission), stomatal closure and leaf senescence, might be explained by a hormonal imbalance. Abscission of fruits can be promoted by ethylene [16, 33] and it is well documented that stomatal closure can be induced by abscisic acid (ABA) [9, 18, 37]. Leaf senescence can be promoted by ethylene [1, 2, 21] and ABA [28].

Initial observations of Dabek and Hunt [7] indirectly suggested changes in PGR in LY-affected plants. They found that treatment with gibberellic acid and kinetin (if added with indoleacetic acid) caused re-greening of detached yellowed pinnae. More recently, results from our laboratory showed that treatment of symptomless coconut palms with the ethylene-releasing agent ethephon promoted nutfall, and as in LY-affected palms, most nuts fell within a few days regardless of their developmental stage [20], suggesting that ethylene promotes LY-associated nutfall. In order to explain LY-induced stomatal closure, leaf ABA content was measured in diseased palms. It increased, but this increase started when stomatal conductance was strongly reduced [20]. Neither can LY-induced stomatal closure be explained by changes in the water status of the leaves, since, as mentioned previously, it remained unchanged in LY-affected palms during the early and intermediate stages of the disease [20]. Further research is needed to determine what induces permanent stomatal closure in LY-affected palms. Leaf senescence, on the other hand, occurred concurrently with the increase in leaf ABA concentration [20]. An increase of ethylene forming enzyme activity was observed before leaf senescence and treatment of symptomless palms with ethephon promoted leaf senescence [20]. These results suggest that leaf senescence in LY-affected palms may be the result of a hormonal imbalance.

It is interesting that LY, as in other yellows-type diseases [22], promotes stomatal closure and that this occurs before the onset of leaf senescence. Thimann and

Satler [35, 36] proposed that stomatal closure triggers leaf senescence. They found that treatments that promoted stomatal closure, including that simulating stomatal closure by covering the leaf surface with a film of Vaseline, promoted senescence [35] and treatments that resulted in stomatal opening delayed senescence [36]. They proposed that the effects of stomatal closure might not cause senescence itself, rather it might lead to an increase of some internal factor such as ABA or ethylene [36]. Whether stomatal closure promotes leaf senescence in LY-affected palms cannot be ascertained presently, but it is a hypothesis warrants testing.

3. Biochemical changes in roots

Eden-Green [11] and Eden-Green and Waters [12] reported a correlation between foliar yellowing and root necrosis in LY-affected palms. Eskafi *et al.* [14], also found extensive root necrosis in diseased palms. Recent studies in our laboratory have also showed biochemical abnormalities in roots (both primary and secondary) of LY-affected palms including sharp decreases in respiration, total sugars and reducing sugars [17]. These decreases started before symptom appearance. In the case of primary roots, the decreases in total and reducing sugar contents were followed by later increases.

In contrast with mature leaves, MLOs have been readily found by DNA probing in root tissues in both symptomatic and positive symptomless palms (Escamilla *et al.*, this volume). Therefore, the biochemical changes observed in roots are not necessarily the result of indirect MLO action as proposed for mature leaves.

It is known that PGR metabolism is altered in roots stressed as a result of water deficit [8], or flooding [31]. In these abiotic stresses, there is a decreased production of some PGRs (cytokinins and gibberellins) and an increased production of others (ABA and the ethylene precursor ACC). It has been proposed that these changes in the root-PGRs can mediate responses in shoots such as stomatal closure, leaf growth [8] and leaf senescence [31]. In LY-affected palms, decreased root respiration and increased root necrosis occurs before the development of any symptom in the shoots. It is possible that root damage in LY-affected palms might cause an imbalance in root PGR similar to what has been proposed for abiotic stresses, which might in turn trigger changes in aerial parts leading to the development of LY symptoms [8, 31].

4. Conclusions

Coconut palms affected by LY show biochemical and physiological abnormalities in shoots and roots. One of these abnormalities, an extensive and permanent stomatal closure, may play an important role in the mode of pathogenesis of the LY-MLO, as it has been proposed for MLOs causing other yellows [22]. In mature leaves, these abnormalities appear to be indirectly promoted since MLO are rarely found in these plant parts. These changes may be mediated by metabolites, such as PGRs, formed elsewhere. Current data supports the occurrance of a PGR imbalance and suggest that some symptoms (nutfall and leaf senescence) may be mediated by such an imbalance. In contrast, changes occurring in roots are not necessarily promoted indirectly since MLO can be readily found in this plant part in positive symptomless or symptom-showing palms. It is possible that root damage in LY-affected palms results in changes in root-PGR metabolism and /or translocation to the shoots, which might result in an imbalance in the shoot-PGR and that this would, in turn, mediate the development of LY-symptoms. This hypothesis should be tested.

Acknowledgments.

We thank M.A. Villanueva (CICY, México) for allowing to quote his unpublished data.

References

1. Aharoni N (1989) Interrelationship between ethylene and growth regulators in senescence of lettuce leaf discs. J Plant Growth Regul 8:309-317
2. Alejar AA, Viser RD and Spencer MS (1988) Ethylene production by attached leaves or intact shoots of tobacco cultivars differing in their speed of yellowing during curing. Plant Physiol 88:329-332
3. Beakbane AB, Slater CHW and Posnette AF (1972) Mycoplasmas in the phloem of coconut, *Cocos nucifera* L., with lethal yellowing disease. J Hort Sci 47:265
4. Cardeña R, Villanueva M, Santamaría J and Oropeza C (1991) Presence of mycoplasma-like organisms in *Cocos nucifera* palms showing lethal yellowing disease symptoms, in Yucatán. Can J Plant Path 13:135-138
5. Carter W (Ed) (1965) FAO lethal yellowing disease project: a compilation of data derived during the FAO mission from June 1962 to June 1965. Coconut Industry Board, Kingston, Jamaica
6. Dabek A (1973) Some physical and anatomical saspects of the lethal yellowing disease of coconut. PhD thesis, University of the West Indies, Kingston, Jamaica
7. Dabek AJ and Hunt P (1976) Biochemistry of leaf senescence in conconut lethal yellowing, a disease associated with mycoplasma-like organisms. Trop Agric 53:115-123

8. Davies WJ and Zhang J (1991) Root signals and the regulation of growth and development of plants in drying soil. Annu Rev Plant Physiol Mol Biol 42:55-76
9. Davies WJ and Jones HG (1991) Abscisic acid, physiology and biochemistry. Oxford: Oxford Bios Scientific
10. Eden-Green SJ (1976) Disease scoring scale for lethal yellowing. In: 16th Report Research Department Coconut Industry Board. Jamaica, pp. 39-41
11. Eden-Green SJ (1976) Root symptoms in coconut affected by lethal yellowing disease in Jamaica. FAO Plant Protect Bull 24:119-122
12. Eden-Green S and Waters H (1982) Collection and properties of phloem sap from healthy and LY-diseased coconut palms in Jamaica. Phytopath 72:667-672
13. Ellis R (1978) Chloroplast proteins and their synthesis. In: Norton G (Ed) Plant proteins, pp 25-40. London: Butterworths
14. Eskafi F, Basham H and McCoy R (1986) Decreased water transport in lethal yellowing-diseased coconut palms. Trop Agric 63:225-228
15. Heinze K, Petzold H and Marwitz R (1972) Beitrag zür atiologie der todlichen Vergilbung der Kokospalme (lethal yellowing disease of coconut palm). Phytopath 34:307-312
16. Henderson J and Osborne DJ (1994) Inter-tissue signalling during the two-phase abscission in oil palm fruit. J Exp Bot 45:943-951
17. Islas-Flores I and Oropeza C (1995) Biochemical changes in roots of lethal yellowing-affected coconut palms. Submitted to Phytopathology
18. Jones RJ and Mansfield TA (1970). Suppression of stomatal opening in leaves treated with abscisic acid. J Exp Bot 21:714-719
19. Kollar A, Seemüller E and Krczal (1989) Impairment of the sieve tube sealing mechanism of trees affected by mycoplasma-like organisms. J Phytopath 124:7-12
20. León R, Alpizar L, Escamilla A, Santamaria J and Oropeza C (1995) Physiological and biochemical changes in shoots of lethal yellowing-affected coconut palms. Can J Plant Path (submitted)
21. Matoo AK and Aharoni N (1988) Ethylene and plant senescence. In: LD Nooden and AC Leopold (Eds) Senescence and aging in plants, pp 241-280 San Diego: Academic Press
22. Matteoni JA and Sinclair WA 1983. Stomatal closure in plants infected with mycoplasmalike organisms. Phytopath 73:98-402
23. McCoy RE (1972) Remission of lethal yellowing in coconut palms treated with tetracycline antibiotics. Plant Dis Rptr 56:1019-1021
24. McCoy RE (1976) Uptake, translocation and persistence of oxytetracycline in coconut palm. Phytopath 66:1038-1042
25. McCoy RE, Howard FW, Tsai JH, Donselman HM, Thomas DL, Basham HG, Atilano RA, Eskafi FM Britt L and Collins ME (1983) Lethal yellowing of palms. Univ Florida Agric Exp Sta Bull No 834
26. McDonough J and Zimmermann J (1979) Effect of lethal yellowing on xylem pressure in coconut palms. Principes 66:1038-1042
27. Nienhaus F, Schuiling M, Gliem G and Spittel A (1982) Investigations of the aetiology of the lethal disease of coconut palms in Tanzania. Z Pflanzensch 89:185-193
28. Nooden LD (1988) Abscisic acid, auxin and other regulators of senescence. In: LD Nooden and AC Leopold (Eds). Senescence and aging in plants, pp 329-368. San Diego: Academic Press
29. Oropeza CM, Santamaría JM, Villanueva MA, and Loyola-Vargas VM (1991) Physiology and biochemistry of lethal yellowing in *Cocos nucifera*. Principes 35:208-218
30. Plavsic-Banjac B, Hunt P and Maramorosh K (1972) Mycoplasma-like bodies associated with lethal yellowing disease of coconut palms. Phytopath 62:298-299
31. Reid D and Bradford H (1984) Effect of flooding on hormone relations in flooding and plant growth. In: TT Koslowski (Ed) Flooding and plant growth, pp 195-219. London: Academic Press
32. Sánchez G (1991) Analysis of free arginine in palms affected by lethal yellowing. B.Sc. thesis. Universidad Autónoma de Yucatán, Mérida, México

33. Sexton R, Lewis A, Trewavas A and Kelly P (1985) Ethylene and abscission. In: JA Roberts and GA Tucker (Eds). Ethylene and plant development, pp 173-176. London: Butterworths
34. Stemmner W, Archer B, Daniels J, Davies C and Eden-Green S (1982) Effect of lethal yellowing on the composition of the phloem sap from coconut palms in Jamaica. Phytopath 72:672-675
35. Thimann KV and Satler SO (1979) Relation between leaf senescence and stomatal closure: senescence in light. Proc Natl Acad Sci USA 76:2295-2298
36. Thimann KV and Satler SO (1979) Relation between leaf senescence and stomatal opening: senescence in darkness. Proc Natl Acad Sci USA 76:2770-2773
37. Zhang J, Schurr U and Davies WJ (1987) Control of stomatal behaviour by abscisic acid which apparently originates in the roots. J Exp Bot 38:1174-1181
38. Zimmerman M (1979) Mycoplasma diseases and long distance transport in plants. NSC Symposium Series, pp. 37-42

Detection and diagnosis of lethal yellowing: conventional methods and molecular techniques

NIGEL A. HARRISON, PATRICIA A. RICHARDSON and JIM H. TSAI
University of Florida, Fort Lauderdale Research and Education Center, 3205 College Avenue, Fort Lauderdale FL 33314, USA

1. Introduction

Lethal yellowing (LY) is a fast spreading, highly destructive disease of coconut palms and 33 additional palm species. The disease presently occurs in several countries of the western Caribbean region, where it has virtually eliminated the susceptible Atlantic Tall ecotype from many localities. A mycoplasmalike organism (MLO) is believed to be the causal agent of LY. The LY MLO, like all other plant pathogenic MLOs, remains unculturable and thus precludes culture techniques for disease diagnosis. Diagnosis of LY has relied primarily upon symptom identification. Unequivocal disease diagnosis requires further visual confirmation of the presence of MLOs in affected palm tissues. This has traditionally been accomplished by either electron microscopy or DAPI staining in conjunction with epifluorescence microscopy. Although quite sensitive, both techniques are non-specific since they cannot distinguish one MLO from another. DNA-based techniques have recently been developed which enable both sensitive and specific detection of the LY MLO. Use of these new techniques as well as conventional methods for diagnosis of LY are discussed.

2. Conventional methods

2.1 Symptoms

The symptoms of LY on coconut palm (*Cocos nucifera* L.), Atlantic Tall, are diagnostic for the disease. They chronologically include premature shedding of all

C. Oropeza et al. (eds.), Lethal Yellowing: Research and Practical Aspects, 79–91.

fruit on the palm, discoloration (blackening) and death of emerged and unemerged flower spikelets (inflorescences), yellowing of older and then younger leaves, root necrosis, collapse of the newly emerged leaf (spear) and soft rot within the crown that invariably precedes death of the palm. Infected trees usually die within 3 to 6 months after the appearance of the first symptoms.

No single symptom is diagnostic for LY disease. Premature nutfall and foliar discoloration may be induced by abiotic factors such as nutritional deficiencies. In their advanced stages, coconut lethal declines caused by *Phytophthora* bud rot, red-ring and hartrot diseases, or bud infestations by palm weevils, share a number of symptoms in common with LY and can be confused with it. The appearance and characteristic progression of symptoms (syndrome development) are what accurately identify LY disease.

Field diagnosis of LY may also be complicated by other factors. For example, non-bearing palms lack fruit and flower symptoms. Foliar discoloration also varies among coconut ecotypes. For most tall-type coconut palms with LY, leaves turn a golden yellow before dying, while the leaves of dwarf-type palms generally turn a reddish to greyish-brown. In southern Florida, where a diversity of palms are grown, LY also affects 33 palm species besides coconut palm (Table 1). On these other species, early stage (primary) symptoms (premature fruit drop and inflorescence necrosis) are essentially the same as those of coconut. However, the stage at which spear collapse occurs differs for individual species. For example, in date palms (*Phoenix dactylifera*), this is usually the first visible foliar symptom of disease. Leaf yellowing beginning with the lower fronds is typical in susceptible *Caryota, Livistona*, and *Pritchardia* species. A browning rather than yellowing of leaves is associated with the appearance of LY on *Veitchia* and *Phoenix* spp.. Since frond discoloration is not nearly as dramatic on other palms as on coconut, the development of early stage symptoms on these other species is frequently overlooked [13].

2.2 Electron microscopy

Mycoplasmalike organisms (MLOs) are cell wall-less prokaryotes that multiply and become systemically distributed within, but restricted to, phloem tissues of their respective plant hosts. These organisms are too small to be resolved by light microscopy. By transmission electron microscopy (TEM) of ultrathin sections, MLOs appear to consist of rounded to filamentous bodies containing granules the size of prokaryotic ribosomes and strands of DNA that apparently condense during specimen preparation (Fig. 1). In sieve tube elements of palms, cells of the LY MLO are generally 142-295 nm in diameter and may vary from 1-16 μm in length [20, 23].

Table 1. Palm species identified as hosts of lethal yellowing disease in southern Florida.

Species	Common name	Species	Common name
Aiphanes lindeniana	-	*Nannorrhops ritchiana*	Mazari palm
Allagoptera arenaria	Seashore palm	*Neodypsis decaryi*	Triangle palm
Arenga engleri *	Dwarf sugar palm	*Phoenix canariensis*	Canary Islands date palm
Borassus flabellifer	Palmyra palm	*Phoenix dactylifera* *	Date palm
Caryota mitis *	Clustering fishtail palm	*Phoenix reclinata*	Senegal date palm
Caryota rumphiana *	Giant fishtail palm	*Phoenix sylvestris* *	Silver date palm
Chelyocarpus chuco *	-	*Pritchardia affinis*	Kona palm
Chrysalidocarpus cabadae	Cabada palm	*Pritchardia pacifica*	Fiji fan palm
Cocos nucifera *	Coconut palm	*Pritchardia remota*	-
Corypha elata	Gebang palm	*Pritchardia thurstonii*	Thurston palm
Dictyosperma album *	Princess palm	*Ravenea hildebrantii*	-
Gaussia attenuata	Puerto Rican gaussia palm	*Syagrus schizophylla*	Arikury palm
Howea belmoreana	Belmore sentry palm	*Tranchycarpus fortunei**	Windmill palm
Hyophorbe verschaffeltii *	Spindle palm	*Veitchia merrillii* *	Manila palm
Latania spp.	Latan palms	*Veitchia macdanielsii*	Sunshine palm
Livistonia chinensis *	Chinese fan palm	*Veitchia montgomeryana*	Montgomery palm
Livistonia rotundifolia *	Footstool palm	*Veitchia arecina* *	Arecina palm

* Identity of LY MLO associated with this diseased palm species has been verified by DNA probe usage and LY MLO-specifc PCR.

The first observations of an MLO association with LY-diseased coconut palms were obtained by TEM [10]. Verification of MLO presence in palm tissues by this method has since provided an important means of confirming questionable LY diagnoses. From extensive TEM surveys of various LY-affected species, it is apparent that overall MLO concentrations are quite low and unpredictably localised in tissues of most symptomatic palms. The pathogen can be most readily found in the young phloem-rich leaf bases surrounding the apical meristem and, to a lesser extent, in inflorescences and tertiary roots. However, in most mature tissues, MLO concentrations are generally below detectable levels.

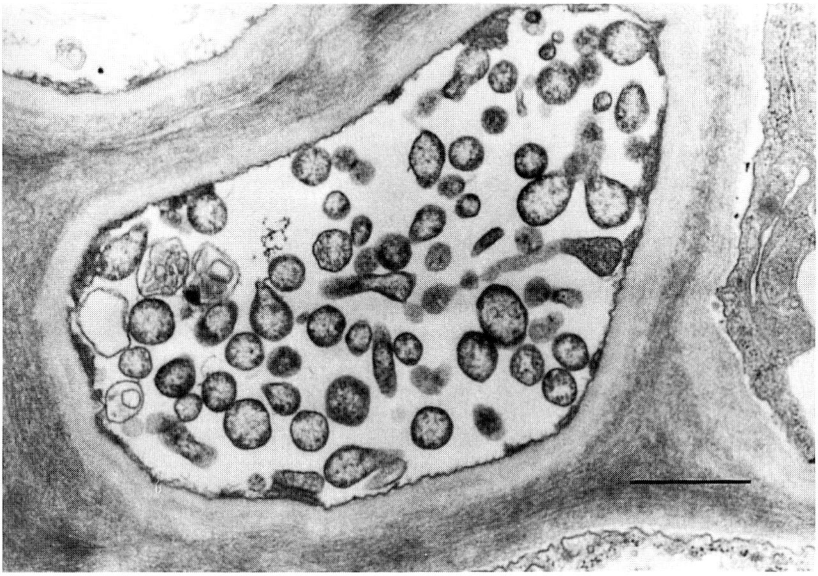

Fig. 1. Electron micrograph of a transverse ultrathin section through a phloem sieve element of a diseased plant containing MLOs. Scale bar 1.0 μm.

2.3 DAPI fluorescence test

MLO infections of plants are characterised by an associated accumulation of DNA within phloem sieve tube elements. Presence of these accumulations can be demonstrated by treatment of either fresh or chemically preserved tissues with DAPI (4'-6' diamidino-2-phenylindole), a reagent that binds to DNA and fluoresces under UV radiation [15]. For detection of low titre MLO infections of woody host plants, fixed and stored tissues have proved superior to use of fresh tissues alone. Tissues are fixed and stored in 2.5% glutaraldehyde in 0.1 M PO_4 buffer, pH 7, and refrigerated at 4°C for several days prior to examination by the DAPI fluorescence test. Optimal staining of tangential or longitudinal sections of nominal thickness (*ca.* 15 μm) from preserved specimens may be achieved by first bathing sections in DAPI solution (2-4 μg ml^{-1} DAPI in 0.1 M PO_4 buffer, pH 7) for 30 min. and then mounting sections on microscope slides in 50% clear corn syrup [19]. MLO cells and nuclei in phloem sieve elements emit a blue-white fluorescence when examined by epifluorescence microscopy using an exciter filter allowing peak transmission at 360 nm and a barrier filter with a short-wave cut-off at 420 nm. Normal sieve tubes are devoid of fluorescence and are usually invisible.

Although the DAPI fluorescence test has seen only limited application for histochemical diagnoses of LY, this non-specific detection technique has been used with considerable success for confirming MLO infections of Tanzanian coconut palms with lethal disease [4, 17]. Palm tissues most suited for MLO detection by TEM are also the tissues of choice for investigation by DAPI staining.

3. Molecular techniques

3.1 DNA probes

Improvements in methods for extraction of MLOs from host tissues, coupled with the advent of recombinant DNA technology, have made the cloning of MLO DNA possible [11]. Cloned anonymous fragments of LY MLO DNA used as molecular probes in DNA hybridisation assays enable sensitive and specific detection of the pathogen and provide a ready means to quantify MLO populations in infected palm tissues [6, 7]. Some cloned probes, presumably representing highly conserved LY MLO chromosomal DNA sequences, detect a range of different MLOs (MLO-specific probes), whereas others appear to detect only the LY MLO (pathogen-specific probes). All of the MLO-specific probes characterised thus far, contain reiterative genomic DNA sequences. They produce stronger hybridisation signals than do the LY MLO-specific probes, which consist of only low copy genomic DNA sequences. For example, the level of detection sensitivity achieved with MLO-specific probe LYI43 (2.3-kbp) is approximately 8-fold greater than that of LY MLO-specific probe LYTC24 (4.4-kbp), despite the smaller size of the former probe (Fig. 2).

In contrast to the conventional methods of TEM and DAPI staining for confirmation of LY diagnoses, DNA-based diagnostic methods are not constrained by specimen size or the quantity of palm tissue which can be readily examined for MLO presence. Extraction techniques have been developed that incorporate use of osmotically-supplemented buffers and differential centrifugation to obtain MLO-enriched preparations from relatively large quantities (50-200 g) of fresh plant tissues (*e.g.* entire palm hearts) [6, 18]. While such large-scale extractions and associated MLO-enrichment ensure that final preparations contain detectable quantities of LY MLO DNA, this approach to sample preparation for screening with DNA probes is probably impractical for all but experimental purposes.

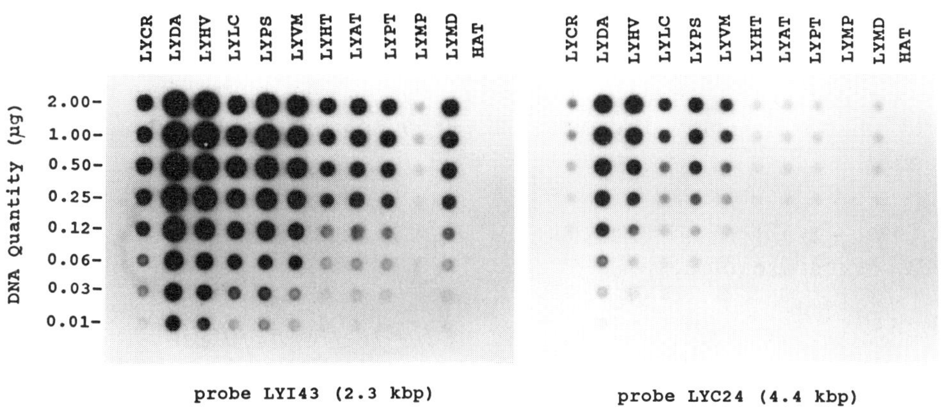

Fig. 2. Dot hybridisation detection of the lethal yellowing (LY) mycoplasmalike organism in various palm species displaying mid-stage foliar symptoms of LY disease. Duplicate blots of DNA derived from immature leaf bases of palms were screened with: (A) MLO-specific DNA probe LY143 (2.3 kbp), or (B) LY MLO-specific probe LYTC24 (4.4 kbp). LY-affected palms: LYCR, *Caryota rumphiana*; LYDA, *Dictyosperma album*; LYHV, *Hyophorbe verschafeltii*; LYLC, *Livistona chinensis*; LYPS, *Phoenix sylvestris*; LYVM, *Veitchia merrillii*; *Cocos nucifera* ecotype: LYHT, Hawaiian Tall; LYPT, Panama Tall; LYAT, Atlantic Tall; LYMP, Maypan hybrid; LYMD, Malayan Green Dwarf; HAT, healthy Atlantic Tall.

Small-scale extractions of total DNA from 1-3 g of palm tissue are quickly and inexpensively obtained by using a standard CTAB protocol [5]. Modification of the DNA extraction buffer to include 1-2% mercaptoethanol and 1% polyvinylpyrrolidone (PVP-40) can greatly improve the quality of DNA from tissues such as roots and leaves which contain elevated levels of phenolic compounds. DNA derived from fresh tissues ground directly in extraction buffer is usually inferior in both quality and yield to DNA from tissues initially ground in liquid nitrogen. Also, final DNA yields may be influenced by the age and type of tissue used for extraction. However, direct processing of fresh tissues may facilitate the screening of large numbers of palm samples.

Immature phloem-rich leaf bases, considered the richest source of the LY agent in diseased palms, are ideal samples for small-scale extractions. Although variable, MLO concentrations in leaf bases appear to be more uniform among individual palms of the same species than between species. For example, in silver date palm (*Phoenix sylvestris*) and windmill palm (*Trachycarpus fortunei*), MLO concentrations revealed by dot hybridisations were at least 10- to 12-fold greater than those usually encountered in tall coconut palms at a comparable stage of disease development (Fig. 2).

Probe-based surveys of other palm tissues have also revealed consistently detectable, although lower, concentrations of the LY MLO in flower spikelet and rachis tissues of partially necrotic inflorescences. However, detection of the LY MLO within randomly sampled roots and leaves has been considerably less reliable and indicative of an uneven distribution of the pathogen throughout these organs. Also, only weak probe hybridisation signals have been associated with most positive detections in coconut leaves, a result that concurs with earlier TEM observations [20, 22] of exceptionally low MLO titres in these tissues.

Use of diagnostic DNA probes in a search for additional plant hosts of LY disease in Florida, has established that the MLO associated with decline disease of the arborescent monocot *Pandanus utilis* [21] is, in fact, the LY MLO. This finding extends the known host range of LY beyond the Arecaceae and validates the concept of possible non-palm hosts contributing to the epidemiology of LY as alternative reservoirs for pathogen increase and dispersal.

Detection of the LY MLO within the cixiid planthopper *Myndus crudus*, the putative vector of the pathogen [9], remains to be convincingly demonstrated by DNA probe hybridisation. With the possible exception of two weak positives, all DNA samples derived from the remainder of the 5 000 individuals screened so far have tested negative for the presence of LY MLO. Native adult *M. crudus* examined in this manner were collected mostly from foliage of non-symptomatic palms on the grounds of the University of Florida's Fort Lauderdale Research and Education Center during 1990-1992; an interval that also coincided with a low incidence of LY among the palms at this particular site. Insect surveys for the purpose of confirming the vector(s) of LY are on-going and are now limited to insects captured from diseased palms only.

3.2 Polymerase chain reaction

Polymerase chain reaction (PCR) involves a primer-directed amplification of target DNA using a thermostable DNA polymerase [16]. Selective amplification of

MLO DNA from mixtures with host DNA by PCR represents the newest approach for detection and identification of MLOs. Several oligonucleotide primers based upon conserved mollicute-specific and MLO-specific ribosomal (rRNA) gene sequences permit universal amplification of partial to full-length MLO 16S rRNA genes during PCR [1, 3, 12]. Thus far, two primer pairs 5'-AAG AGT TTG ATC CTG GCT CAG GAT T-3' (forward), 5'-TTG TAG GGA TAC CTT GTT ACG ACT TA-3' (reverse) [3] and 5'-ACG ACT GCT GCT AAG ACT GG-3' (forward), 5'-TGA CGG GCG GTG TGT ACA AAC CCC G-3' (reverse) [12], have consistently amplified rDNA of the LY MLO when template DNAs for PCR are derived from either heart or inflorescence tissues of symptomatic palms [8, 12].

While PCR incorporating these primer pairs provides an exceptionally sensitive technique for detection of MLO infection of plants, the size of the resulting rDNA amplimer is comparable for all MLOs. Further restriction site analysis or cycle sequencing of the rDNA amplimer is necessary for MLO identification. Fragment patterns revealed by agarose gel electrophoresis of MLO rDNA after digestion with the restriction endonucleases *Alu*I, *Rsa*I, or *Mse*I are particularly useful for distinguishing the LY MLO from other MLOs [12]. *Hinf*I or *Taq*I digests also discriminates rDNA of the Caribbean LY MLO from related MLOs associated with African coconut lethal decline diseases [7].

PCR primer sets derived from MLO ribosomal gene sequences have recently been reported that impart a greater exclusivity to MLO detection by PCR [2]. This new development indicates that it may also be feasible that a similar group-specific detection of the LY MLO and related MLOs may arise through use of primer sets that exploit sequences from variable regions of the LY MLO rRNA gene operon. To date, pathogen-specific detection by PCR using MLO rRNA primers has not yet been demonstrated.

Development of a PCR assay enabling specific detection of the LY MLO has recently been accomplished by selecting a pair of primers, 5'-CAT ATT TTA TTT CTT TTG CAA TCT G-3' (forward) and 5'-TCG TTT TGA TAA TCT TTC ATT TGA C-3' (reverse), from disease-specific probe LYTC24, a cloned 4.4-kb genomic DNA sequence unique to the LY MLO [8]. This particular primer set was chosen to permit PCR amplification of a DNA product of sufficient size (1.0-kb) that could be readily resolved by electrophoresis in standard agarose gels.

For performance of LY MLO-specific PCR, a standard 50 μl reaction mixture typically contains the following components: 25-50 ng of palm sample DNA as template, 50 ng of each primer, 125 μM of each dNTP, and 1.5 U of *Taq* DNA polymerase with standard PCR buffer [10]. Reaction mixtures are subjected to the following thermocycling conditions: denaturation at 94°C for 30 s (90 s for first cycle), annealing at 53°C for 50 s and primer extension at 72°C for 80 s (10 min for final cycle). Products from 10 μl of each final reaction mixture are separated by 1% agarose gel electrophoresis and then visualised by UV transillumination after staining the gel with an aqueous 1 μg ml^{-1} ethidium bromide solution.

The reliability of the pathogen-specific primer set for detection of the LY MLO has been established by repeated testing of template DNAs extracted from hearts or unemerged and newly emergent inflorescences of coconut and 11 additional palm species, each species displaying early- to mid-stage LY foliar symptoms. Exclusive amplification of a 1.0-kb LY MLO DNA product was achieved for all symptomatic palms, which included 14 coconut palms (four ecotypes and one hybrid) and 17 additional palms. No product was amplified from DNA samples derived from heart tissues of young containerised palms grown in shade-houses and used for comparative purposes. Furthermore, positive detection of the LY MLO in heart tissue of coconut palm was achieved, in some instances, by using as little as 5 *pg* of DNA as template and 35 cycles of PCR. In terms of sensitivity, PCR increased the detectability of the LY MLO 100 000 times more than LY-specific DNA probes.

The capability of PCR to detect low MLO concentrations has made investigation of LY MLO distribution in palm foliage possible. Forty cycle reactions were necessary to reliably detect presence of the pathogen in mostly non-symptomatic leaves of coconut palms with earliest visible symptoms of disease. In three non-bearing palms examined in detail, detectable MLO concentrations were limited to the younger leaves of the crown when test samples consisted of leaflets removed from basal portions of leaves. In further limited sampling of 10 coconut, one Manila (*Veitchia merrillii*) and one giant fishtail palm (*Caryota rumphiana*), detection success was 100% with leaflet samples removed at random from the mid- to lower-portions of the spear, the most immature leaf.

Pathogen-specific PCR has also revealed LY MLO infection of seed from three coconut palms exhibiting primary stage symptoms. DNAs derived from five of 34 excised embryos clearly tested positive for MLO DNA following 40 cycles of PCR. At least one MLO-positive embryo was detected among seed from each palm (Fig. 3). For two palms, a Panama Tall and a Malayan Green Dwarf, samples consisted of embryos collected from fallen nuts only. All embryo samples from a third palm, also a Malayan Green Dwarf, were removed solely from seed that remained attached to the palm. Since the viability of the infected embryos was not ascertained during this initial study, further work is needed to determine the significance of this result.

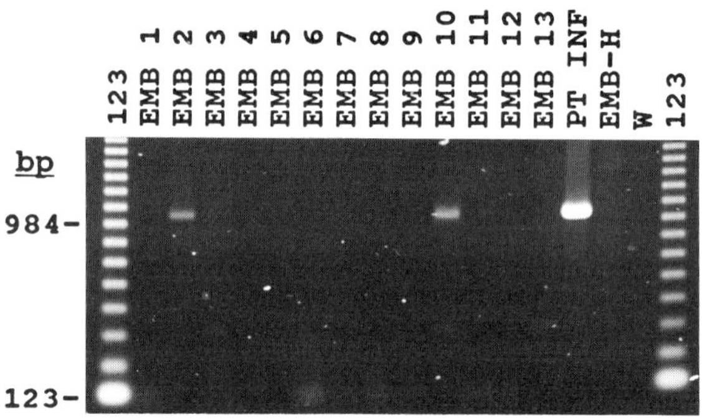

Fig. 3. Pathogen-specific polymerase chain reaction (PCR) detection of the lethal yellowing (LY) mycoplasmalike organism in embryos from seed of a *Cocos nucifera* Panama Tall (PT) with primary stage symptoms of LY disease. PCR products in one fifth of each final reaction mixture were analysed by 1% agarose gel electrophoresis. EMB 1-13, individual embryos from an LY-affected PT coconut palm; PT INF, inflorescence from LY-affected PT coconut palm; EMB-H, embryo from symptomless (healthy) Malayan Green Dwarf coconut palm; W, water control. 123, BRL 123-bp ladder.

Acquisition of the LY MLO by a phloem-feeding Homopteran from a non-bearing Atlantic Tall coconut palm with mid-stage LY disease symptoms has recently been investigated by PCR. In an initial experiment, 31 of 60 adult non-vector corn planthoppers (*Peregrinus maidis* Ashmead), reared from eggs, tested positive for presence of the LY MLO when subjected to a 3-day acquisition access period (AAP) on symptomatic coconut leaves. In two subsequent experiments, a substantial increase in mortality occurred among corn planthoppers during 2 or 3 day AAPs on LY-diseased Malayan Green Dwarf and Hawaiian Tall. These losses also coincided with a reduction in the proportion of MLO-positive individuals (<10%) reclaimed from each palm.

An association of the LY MLO with *M. crudus* planthoppers has also been demonstrated for the first time by pathogen-specific PCR (Fig. 4). Using as little as 1/50th of the total DNA extracted from each planthopper as template, a

40-cycle reaction was sufficient to readily detect the LY agent in 36 of 94 individuals collected solely from foliage of a non-bearing *C. nucifera,* Atlantic Tall, with mid-stage foliar yellowing symptoms.

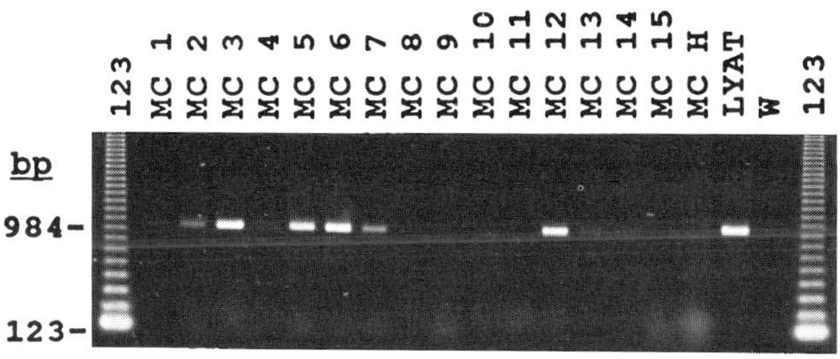

Fig. 4. Pathogen-specific polymerase chain reaction (PCR) detection of the lethal yellowing (LY) mycoplasmalike organism in the cixiid planthopper *Myndus crudus* captured from foliage of a non-bearing *Cocos nucifera* Atlantic Tall (AT) palm with mid-stage foliar yellowing symptoms. MC 1-15, PCR products from indiviual planthoppers; MC H, healthy egg-reared planthopper; LYAT, spear leaf from LY-affected AT coconut palm; W, water control. 123, BRL 123-bp ladder.

4. Conclusions

Diagnostic DNA probes have greatly improved detection and identification of the MLO associated with LY disease. However, because of the unusually low titre of the pathogen in most palm tissues, use of probes has required that they be labelled with a radioisotope to consistently obtain the necessary detection sensitivity. This prerequisite has limited their use for investigations of LY elsewhere in the Caribbean, where the disease is presently most active.

The convenience of PCR makes it attractive for LY research. Also, the tremendous sensitivity of LY MLO-specific PCR makes it possible to develop practical sampling strategies for routine indexing of palms for infection by this pathogen. As such, it has relevance for service-oriented programs such as those dealing with plant health and plant improvement. Furthermore, the capability to now monitor insect acquisition of the LY MLO should open up new avenues of LY research, allowing detailed studies of vector biology, host resistance, and searches for alternative plant hosts, which have previously been beyond the reach of the available technology.

Acknowledgements

This work was funded by the USDA Tropical/Subtropical Agriculture Special Grants Program (Grant No. 90-34135-5165 and 92-34135-7281). We are grateful to Jack Kramer for assistance with DNA sequence analysis and oligoprimer design.

References

1. Ahrens U and Seemüller E (1992) Detection of DNA of plant pathogenic mycoplasmalike organisms by a polymerase chain reaction that amplifies a sequence of the 16S rRNA gene. Phytopath 82:828-832
2. Davis RE and Lee IM (1993) Cluster-specific polymerase chain reaction amplification of 16S rDNA sequences for detection and identification of mycoplasmalike organisms. Phytopath 83:1008-1011
3. Deng S and Hiruki C (1991) Amplification of 16S rRNA genes from culturable and nonculturable mollicutes. J Microbiol Meth 14:53-61
4. Deutsch E and Nienhaus F (1983) Further studies on the distribution of mycoplasmalike organisms in different tissues of lethal diseased coconut palms in Tanzania. Z Pflkrankh Pflschutz 90:278-284
5. Doyle JJ and Doyle JL (1990) Isolation of plant DNA from fresh tissue. Focus 12:13-15
6. Harrison NA, Bourne CM, Cox RL, Tsai JH and Richardson PH (1992) DNA probes for detection of mycoplasmalike organisms associated with lethal yellowing of palms in Florida. Phytopath 82:216-224
7. Harrison NA, Richardson PA, Jones P, Tymon AM, Eden-Green S and Mpunami AA (1994) Comparative investigation of mycoplasmalike organisms associated with Caribbean and African coconut lethal decline diseases by DNA hybridizations and PCR assays. Plant Dis 78:507-511
8. Harrison NA, Richardson PA, Kramer JB and Tsai JH (1994) Detection of the mycoplasma-like organism associated with lethal yellowing disease of palms in Florida by polymerase chain reaction. Plant Path 43:998-1008
9. Howard FW, Norris RC and Thomas DL (1982) Evidence of transmission of palm lethal yellowing agent by a planthopper, *Myndus crudus* (Homoptera: Cixiidae). Trop Agric 60:168-171
10. Innis MA and Gelfand DH (1990) Optimization of PCRs. In: MA Innis, DH Gelfand, JJ Sninsky and TJ White, eds. PCR Protocols: A Guide to Methods and Applications, pp 3-12. New York: Academic Press
11. Kirkpatrick BC, Stenger DC, Morris TJ and Purcell AH (1987) Cloning and detection of DNA from a nonculturable plant pathogenic mycoplasmalike organism. Science 238:197-200
12. Lee IM, Hammond RW, Davis RE and Gunderson DE 1993 Universal amplification and analysis of pathogen 16S rDNA for classification and identification of mycoplasmalike organisms. Phytopath 67:802-807
13. McCoy RE, Howard FW, Tsai JH, Donselman HM, Thomas DL, Basham HG, Atilano RA, Eskafi FM, Britt L and Collins ME (1983) Lethal yellowing of palms. Univ Florida Agric Expt Sta Bull No 834
14. Plavsic-Banjac B, Hunt P and Maramorosch K (1972) Mycoplasmalike bodies associated with lethal yellowing disease of coconut palms. Phytopath 62:298-299
15. Russel WC, Newman C and Williamson DH (1975) A simple cytochemical technique for demonstration of DNA in cells infected with mycoplasmas and viruses. Nature 253:461-462
16. Saiki RK, Gelfand DH, Stoffel S, Scharf SJ, Higuchi G, Horn T, Mullis KB and Erlich HA (1988) Primer-directed enzymatic amplification of DNA with a thermostable DNA polymerase. Science 239:487-491

17. Schuiling M and Mpunami AA (1990) The lethal disease of coconut in Tanzania: review of research up to date and preliminary results of resistance trials. In: ML Robert and D Zizumbo, eds. La problemistica del amarillamiento letal de cocotero en Mexico, pp 171-183. Mérida: Centro de Investigación Científica de Yucatán
18. Sears BB, Lim P-O, Holland N, Kirkpatrick BC and Klomparens KL (1989) Isolation and characterization of DNA from a mycoplasma-like organism. Mol Plant-Microbe Interact 2:175-180
19. Sinclair WA, Griffiths HM, Davis RE and Lee IM (1992) Detection of ash yellows mycoplasmalike organisms in different tree organs and in chemically preserved specimens by a DNA probe vs. DAPI. Plant Dis 76:154-158
20. Thomas DL (1979) Mycoplasmalike bodies associated with lethal declines of palms in Florida. Phytopath 69:928-934
21. Thomas DL and Donselman HM (1979) Mycoplasmalike bodies and phloem degeneration associated with declining *Pandanus* in Florida. Plant Dis Reptr 63:911-916
22. Thomas DL and Norris RC (1980) The use of the electron microscope for lethal yellowing diagnosis. Proc Florida State Hort Soc 93:196-199
23. Waters H and Hunt P (1980) The *in vivo* three-dimensional form of a plant mycoplasma-like organism by the analysis of serial ultrathin sections. J Gen Microbiol 116:111-131

Practical use of DNA probes for the detection of lethal yellowing of *Cocos nucifera* L. in México

JOSE A. ESCAMILLA[1], NIGEL A. HARRISON[2], HERNAN NUÑEZ[1], LUCELY ALPIZAR[1], IVAN CORDOVA[1], IGNACIO ISLAS-FLORES[1] and CARLOS OROPEZA[1]

[1]*Centro de Investigación Científica de Yucatán. Apartado Postal 87, Cordemex, 97310 Mérida Yuc., México*

[2]*University of Florida, Fort Lauderdale Reasearch and Education Center, 3205 College Avenue, Fort Lauderdale FL 33314, USA*

1. Introduction

Lethal yellowing (LY) is a fatal disease of palms [5], and is believed to be caused by a mycoplasmalike organism (MLO) [1, 4, 6]. Harrison *et al.* [3] isolated and cloned fragments of LY-MLO DNA from palms affected by the disease in Florida and used them as diagnostic probes. These probes have since proved useful for the detection of the LY pathogen in affected palms in the Yucatán Peninsula of México [2]. LY is a disease that presently cannot be controlled, but measures, such as quarantine, may be taken to reduce its rate of spread (Been, this volume). Effective quarantine requires a constant evaluation of the geographic occurrence of the disease. To achieve this, a reliable diagnostic method is required. This paper reports the detection of LY in symptomatic and symptomless coconut palms (*Cocos nucifera* L.) using a DNA probe, and also reports the use of this technique as a tool for monitoring the spread of LY in the Yucatán Peninsula.

2. Materials and methods

2.1 Plant material

Coconut palms, either with or without symptoms indicative of LY [5], were sampled in three areas. Two of the areas were in the Yucatán Peninsula: an LY-endemic area where LY has been present for over 2 years, and an LY-incipient

C. Oropeza et al. (eds.), Lethal Yellowing: Research and Practical Aspects, 93–99.

area where the initial occurrence of disease was reported no more than 6 months prior to sampling. A third area, consisting of the States of Tabasco and Veracruz, was considered to be free of LY. Young leaves, unemerged inflorescences, and root and shoot meristems were analysed within all three areas except the LY-free area, where shoot meristem were not sampled because this practice is destructive. The number of palms analysed was as follows: LY-endemic area, 8 palms with symptoms and 12 palms without symptoms; LY-incipient area, 6 palms with symptoms and 10 palms without symptoms; LY-free area, 21 palms without symptoms. Immediately after collection, samples were wrapped in moist paper and maintained at 4°C in an insulated box for transportation. Once in the laboratory, samples were immersed in liquid nitrogen until DNA extraction was performed.

2.2 Detection of LY

DNA was extracted from samples using the method of Harrison et al. [5]. Aliquots of each denatured DNA extract were then spotted onto a nitrocellulose membrane and dried at 80°C under vacuum for 2 h. The membrane was sealed inside a plastic bag containing hybridisation buffer consisting of 6x SSC (0.9M NaCl, 0.09M sodium citrate), 20% sodium dodecyl sulfate (SDS), Denhart's solution [7], 100 mg ml^{-1} of salmon sperm DNA, and incubated at 65°C overnight. The membrane was then transferred to a second bag of fresh buffer also containing 20 mg ml^{-1} of biotinylated DNA probe LYI43 [3] previously denatured at 90°C for 15 min. After a further overnight incubation at 65°C, the membrane was washed twice with 2x SSC buffer (containing 0.5% SDS) for 30 min, twice with 2x SSC buffer (containing 0.1% SDS) and two final washes with 2x SSC buffer alone. After incubation with avidine (1:1000) at 25°C, the membrane was washed with two changes of 2x SSC buffer for 30 min, and likewise with developing buffer (0.1M Tris, 0.1M NaCl, 0.2M $MgCl_2$, pH 9.5). Colour development was performed using the same buffer supplemented with 0.4 mg ml^{-1} of bromochloro-indolyl phosphate and 0.5 mg ml^{-1} of nitro-blue tetrazolium. Each positive probe hybridisation was indicated by the appearance of a blue spot. Frequency results were analysed using generalised linear models employing a logit link function.

3. Results and discussion

3.1 Detection of LY

To determine which organs of the palms were more convenient for detection of LY-MLO by DNA probe hybridisation, DNA samples were extracted from mature leaves, young leaves, inflorescences, shoot and root meristems of palms, both with and without symptoms, in areas where LY was endemic, incipient or believed to be absent. Preliminary results suggested that the pathogen was not readily detectable in mature leaves. Harrison *et al.* (this volume) also reported evidence indicating that LY-MLO is less abundant or, perhaps, more unevenly distributed in older as compared to younger leaves. While the LY-MLO was not detected in any of the organs of symptomless palms in the disease-free area in this study, positive detection of the pathogen in various organs of both symptomatic and symptomless palms for each of the LY-affected areas was observed (Table 1).

Detection of the LY-MLO in symptomless (incubating) palms present in both LY-endemic and incipient areas, illustrates the potential of DNA probe hybridisation for early LY diagnosis. In all cases, the LY-MLO could be detected in any of the four organs analysed: shoot meristem, inflorescences, young leaves and root meristems (Table 1). However, for each individual palm, the pathogen was not necessarily detected in all of the samples analysed. The proportion of positive samples varied according to the particular tissues tested, however this experiment was not adequately replicated to gain a statistically significant result. Overall, it appeared as if inflorescences were inferior to the use of other organs for pathogen detection.

When probe hybridisation results obtained for shoot meristem, inflorescences, young leaves and root meristems were pooled, positive detection percentages for each palm were generally greater than those obtained for each organ type alone. Pooled values were in the order of 75-85% per palm, with no significant differences among cases. The proportion of positively testing symptomless palms was similar to that of adjacent symptomatic palms (Table 1). This observation is important for quarantine purposes, because, in a coconut grove, once the first palm showing LY symptoms appears, additional presymptomatic (incubating) palms are likely to be present. Furthermore, this result indicates that the use of DNA probes might provide a means to assess the extent of this initial disease infestation. Also, the probability of detection of LY in a particular locality where the disease is just appearing, for instance when there is only one palm showing symptoms, would increase when sampling the palm with symptoms and as well as those around it. Sampling was carried out this way when monitoring LY spread in western Yucatán Peninsula (see below).

Table 1. Detection of LY in different parts of coconut palms with the DNA probe LYI43

Case	Type of area	LY symptoms	positive samples (%)				positive palms (%)	
			SM	I	YL	RM	SM+I+YL+RM	I+YL+RM
1	LY-endemic	yes	57	33	71	33	85	85
2	LY-endemic	no	62	18	45	41	83	75
3	LY-incipient	yes	60	20	60	60	80	60
4	LY-incipient	no	50	37	16	50	75	50
1-4	Average		57	26	48	45		
5	LY-free	no	nd	0	0	0	0	0

SM = shoot meristem, I = inflorescence, YL = young leaf, RM = root meristem, nd = not determined

The shoot meristem is considered a reliable source of the LY-MLO [5], but its removal from the palm for analysis is a destructive sampling technique for coconut and other susceptible palm species. Therefore, the use of palm organs other than the shoot meristem is preferable. If the percentage of positive palms detected by DNA probe hybridisation is recalculated only on the basis of the combined results obtained with inflorescences, young leaves and root meristem data, practically no variation in the proportion of LY-positives was observed for palms in LY-endemic area (Table 1). Therefore, non-destructive sampling of these palms without seemingly compromising the overall frequency of detection could be used. However, with palms from the LY-incipient areas, a trend towards decreased levels of detection was observed when the shoot meristem data was omitted (Table 1). Hence, when assessing palms for LY in such areas it is advantageous to include the shoot meristem in any proposed sampling scheme.

3.2 Monitoring of LY spread

A survey to monitor the spread of LY in western Yucatán Peninsula was conducted using DNA probe hybridisation as a basis for assessing the disease status of coconut palms in eight locations where LY was newly apparent (Fig. 1).

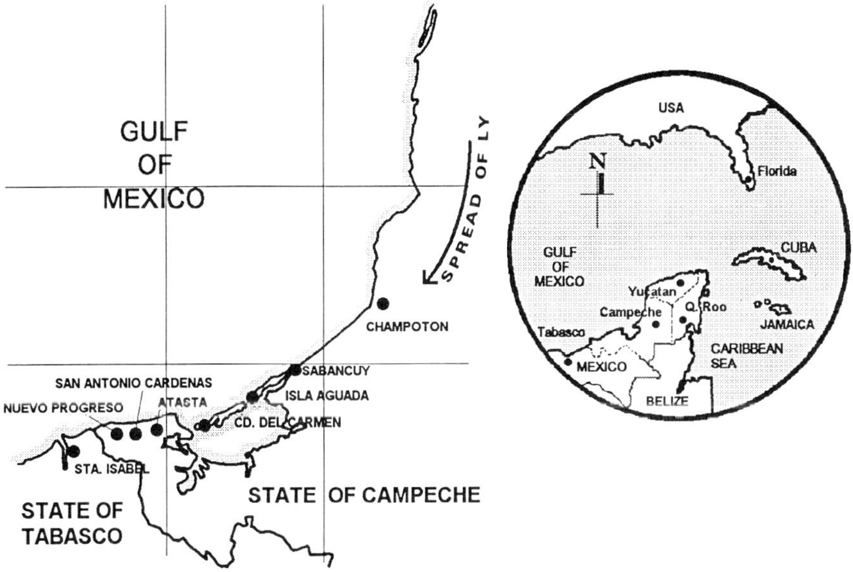

Figure 1. Spread of LY in western Yucatán Peninsula

At each location, young leaves, inflorescences, and shoot and root meristems were collected from each of three or more palms including at least one palm already displaying LY symptoms. When sampling symptomless palms, only those occurring within 10 m of a symptomatic palm were chosen. Although only a limited number of palms were investigated at each location during the course of the survey, presence of LY-infected palms was confirmed at all locations (Table 2). The overall percentage of palms testing positive for infection by the LY-MLO was 70% for symptomatic palms and 66 % for symptomless palms (Table 2).

LY was initially monitored in the State of Campeche (Fig. 1), where it has since moved westward towards the border of the State of Tabasco, the centre of coconut production in eastern México. The last location evaluated in the present study confirmed the disease to be already present within Tabasco. This new development threatens not only coconut production within the state but also endangers production throughout the rest of the country since LY is approaching the Isthmus of Tehuantepec, the narrowest path linking the Gulf of México with the Mexican Pacific coast. The Pacific coast is where the greatest coconut production is centred, and has no barriers to air-borne, vector-mediated dispersal of LY.

Table 2. Monitoring of LY spread with the DNA probe LYI43 in the Yucatán Peninsula

Location	Date of sampling	Palms with LY symptoms		Palms without LY symptoms		Diagnosis of LY per location
		positive / analysed[a]	%	positive / analysed[a]	%	
CAMPECHE						
Champoton	Oct 92	1/1	100	2/2	100	positive
Sabancuy	Oct 92	1/1	100	2/2	100	positive
Isla Aguada	Oct 92	3/3	100	1/1	100	positive
Ciudad del Carmén	Nov 92	2/3	66	1/4	25	positive
Peninsula de Atasta	Jan 93	3/3	100	ns	ns	positive
San Antonio Cardenas	Jan 93	1/3	33	ns	ns	positive
Nuevo Progreso	Nov 92	0/2	0	2/3	66	positive
TABASCO						
Santa Isabel	Jan 93	1/1	100	ns	ns	positive
Total/Average		12/17	70	8/12	66	

[a]Organs analysed: shoot meristem, inflorescences, young leaves and root meristems. (ns) not sampled

Acknowledgements

Monitoring of the spread of LY was carried out in co-ordination with the Mexican phytosanitary authorities: Dirección General de Sanidad Vegetal (SARH) and Sanidad Vegetal del Estado de Campeche (SARH), and the information obtained was used to update the location of the check-points in their quarantine program. This work was partially supported by CONACYT (grant 1270-N9204).

References

1. Beakbane AB, Slater CHW and Posnette AF (1972). Mycoplasmas in the phloem of coconut, *Cocos nucifera* L., with lethal yellowing disease. J Hort Sci 47:265-265
2. Escamilla JA, Harrison NA, Alpizar L and Oropeza C (1993) Detection of lethal yellowing mycoplasmalike organisms by DNA probes in palms of Yucatán, México. In: MK Nair, HH Khan, P Gopalasundaram and EVV Bhaskara Rao, eds. Advances in Coconut Research and Development. pp 597-604. New Dehli: Oxford and IBH New Delhi
3. Harrison NA, Bourne C, Cox R, Tsai J and Richardson P (1992) DNA probe for detection of mycoplasmalike-organisms associated with lethal yellowing of palms of Florida. Phytopath 82:216-224
4. McCoy RE (1973) Effect of various antibiotics on development of lethal yellowing in coconut palm. Proc Florida State Hort Soc 86:503-506

5. McCoy RE, Howard FW, Tsai JH, Donselman HM, Thomas DL, Basham HG, Atilano RA, Eskafi FM Britt L and Collins ME (1983) Lethal yellowing of palms. Univ Florida Agric Exp Sta Bull No 834
6. Plavsic-Banjac B, Hunt P and Maramorosch K (1972) Mycoplasma bodies associated with lethal yellowing disease of coconut palms. Phytopath 62:298-299
7. Sambrook J, Fritsch EF, and Maniatis T (1989) Molecular Cloning. A Laboratory Manual. Vol 1. Cold Spring Harbor: Cold Spring Harbor Laboratory Press

Integrated pest management for the control of lethal yellowing: quarantine, cultural practices and optimal use of hybrids

BASIL O. BEEN
Coconut Industry Board, 18 Waterloo Road, Kingston 10, Jamaica

1. Introduction

Lethal yellowing (LY), the most destructive of diseases affecting the coconut palm (*Cocos nucifera* L.), is perhaps the greatest threat to future coconut cultivation, not only in the Caribbean and the Americas, but worldwide.

Results of research performed on various aspects of the disease have been well documented and will not be repeated in this paper. It is sufficient to note than LY is associated with mycoplasmalike organisms (MLO) [1, 8] which can be transmitted by *Myndus crudus* Van Duzee [4].

2. Control of lethal yellowing

Despite considerable research over many years, no permanent cure has been found for LY. Nevertheless, some control can be achieved by an integrated program making use of resistant planting material, quarantine, chemotherapy, vector control and sanitation.

2.1 Resistant germplasm

Currently, the principal management tool against LY is the use of resistant germplasm. Unfortunately, little is known about the genetic nature of LY resistance. Screening for host plant resistance is difficult because the MLO associated with the disease has been difficult to detect, and has not been cultured *in vitro* or transmitted by artificial means. Using conventional methods, palms that

C. Oropeza et al. (eds.), Lethal Yellowing: Research and Practical Aspects, 101–109.

have survived after years of field exposure to the disease are assumed to be resistant while susceptible palms have been killed. There is the possibility that some survivors may be escapes.

To identify an ecotype with a LY resistance level high enough for commercial use, a number of trials need to be established at many sites where the disease is active or likely to become active. Adequate numbers of palms of known susceptible ecotypes should be included as controls. Regular and frequent inspections are necessary so that losses from LY are not confused with those due to other causes. Conventional screening for LY resistance is time-consuming and therefore there is a need to develop rapid screening methods.

Resistance to LY was first reported in Jamaica by Nutman and Roberts [7] in fields of 'Malayan Red Dwarfs' that has survived despite continuous exposure to the disease. Later, green and yellow fruited 'Malayan Dwarfs' were found to be highly resistant. In a number of replicated field trails established in the 1960s, the three colour forms had a survival rate of 96% [2].

Many coconut ecotypes were introduced into Jamaica during the 1960s and subsequently screened for disease resistance. They varied in LY resistance [2]. Three dwarf varieties from India and Sri Lanka, and the 'King Coconut' have resistance that seems comparable to that of the 'Malayan Dwarfs'. Also of interest, and possible value, are two palms introduced from Cuba in 1951, as well as progeny of the 'Malayan Dwarf' x 'Niu Leka' F_1 hybrids introduced from Fiji in 1933. The original palms from Cuba, and some of their offspring from open pollination, still stand on the original site while the 'Jamaica Talls' established at the same time have long since died from lethal yellowing. Later generations of the original 'Malayan Dwarf' x 'Niu Leka' cross show segregation of characters but possess good a LY resistance rate of 95% survival.

Once resistance has been identified and tested, seednuts of the desired ecotype or hybrid must be produced in commercial quantities for distribution. When the palm is being grown for fruit, it is important that the variety recommended to the farmers has not only good resistance to LY, but also good yield stability, and in hurricane-prone regions, good windstorm resistance.

Although the 'Malayan Dwarf' is more productive than the 'Jamaica Tall' under good conditions [9], it yields poorly under marginal conditions. Due to these yield characteristics, many hybrids were synthesised and widely tested in the hope of discovering LY resistance and tolerance to marginal conditions in one hybrid. This research led to the production of the very successful 'Maypan' F_1 hybrid. The 'Malayan Dwarf' and the 'Maypan' have superseded the 'Jamaica Tall' as the dominant coconut type in Jamaica. Most hybrids with the 'Malayan Dwarf' as one parent have resistance levels between those of both parents, but generally closer to that of the more resistant parent [2].

No case of full recovery from lethal yellowing has been observed, but partial remission of symptoms has been noted. It is not known if resistance is due to tolerance or immunity.

2.2 Quarantine

The movement of living plant material from LY-infected to LY-free areas should be avoided. The following general recommendations have been made by the International Board For Plant Genetic Resources (IBPGR) for the safe movement of coconut germplasm [3]:

a) germplasm should be collected from palms that appear healthy,

b) germplasm should not be moved from sites at which diseases of unknown aetiology occur,

c) germplasm should preferably be moved as embryo cultures or pollen,

d) seednuts may be transferred under certain conditions:

(i) when a thorough pest risk assessment indicated that there are no problems of quarantine concern in the area from which they were collected, or

(ii) from areas where diseases of quarantine concern are present only when embryo culture is not possible, and as long as they are germinated in quarantine.

e) seednuts should never be moved directly from areas where non-cultivable mollicutes or *Phytomonas* occur, to areas not affected by these pathogens,

f) embryos, seedlings and palms from which pollen is collected should be indexed for cadang-cadang and other viroids, as well as for coconut foliar decay virus (CFDV), and

g) the transfer of germplasm should be carefully planned in consultation with quarantine authorities, the relevant indexing laboratory and, when appropriate, the intermediate quarantine facility. The material should be accompanied with the necessary documentations.

The success of quarantine measures depends upon, among other things, the ability and willingness of countries and persons to enforce them and abide by them.

Quarantine measures in Jamaica and Florida comprise legislation against moving living coconut material out of LY areas, but there is little evidence that this has contained the spread of the disease within the island or state. In Jamaica, over 600 seedlings grown from seeds collected from 'Jamaica Talls' with lethal yellowing and planted in two disease-free areas were free from lethal yellowing during a seven-year observation period. Nevertheless, seednuts are not sent from Jamaica to countries that are LY-free and seednuts sent to countries with LY are harvested from disease-free areas because of the possibility of different strains of MLO existing.

2.3 Chemotherapy

Once MLO were found in LY-affected palms, chemotherapy tests were carried out because this group of organisms were known to be sensitive to certain antibiotics. Stem injection of the tetracycline group of antibiotics suppressed symptom development if injected prior to expression of systemic foliar yellowing [5].

In Florida, where palms are grown for aesthetic rather than utilitarian purposes, chemotherapy has been used on a large scale as a partly successful, temporary LY control while resistant palms were being grown. In Jamaica, where the coconut palm is grown commercially mainly for its fruit, chemotherapy was tried but rejected because of high costs and perceived health risks.

2.4 *Control of vector*

Since LY is vector-borne, its spread depends to some extent upon the population density of its vectors. Although the use of chemical sprays was able to reduce levels of *Myndus crudus* populations in Florida [6], the reduction was not large enough for this method to be successfully used in commercial planting. The treatment of all palms around a single diseased palm with systemic insecticides so that vectors would be killed before acquiring the MLO from infected palms, with or without symptoms, has been tried without success [10]. A judicious use of cover-crops could reduce the populations of *Myndus crudus* in commercial coconut plantations because of the nature of its life-cycle.

2.5 *Sanitation*

Felling of LY-affected palms (thereby removing the source of inoculum) has been used as a control measure, but with little success. The main draw-back of the sanitation is that the disease has a long latent period thereby making it impossible, using conventional methods, to diagnose infected palms until months after they have become infected. Sanitation alone is unlikely to bc able to control lethal yellowing, but it may have a part to play when used along with other methods.

3. Coping with lethal yellowing in Jamaica

3.1 *Background*

Lethal yellowing was first observed in the south-western section of Jamaica in 1884, it continued to be endemic in that area for decades and by 1952 had spread over the western half of the island. In 1961, the disease suddenly appeared in the north-eastern section of the island, over 100 km from the nearest case in the west, and then rapidly spread throughout the main coconut growing region. Since its appearance in Jamaica, LY has destroyed millions of palms with correspondingly large economic losses and replanting costs.

3.2 Pre-1961

In the absence of even a rudimentary understanding of the disease, early attempts were made to control its spread by isolation and elimination of outbreaks. Felling and burning failed to halt the disease. In the northwestern sections of Jamaica, the coastal strip had to be replanted at least four times between the turn of the century and 1940.
Government legislation was enacted to control the movement of plant material and soil eastwards into the disease-free area, but the boundary was an open area that was impossible to maintain in accord with any proposed quarantine requirements.

3.3 Post-1961

The outbreak of LY in the main coconut growing region posed an enormous problem for the fledgling Research Department of the Coconut Industry Board. The original research program had to be modified with emphasis being placed on selection and breeding for disease resistance, and the management methods suited to resistant planting materials.

It was realised from the onset, that the additional costs of research on LY could not be provided by the Board alone and external assistance was sought and obtained. Initially, USAID provided scientists and later the Food and Agriculture Organisation (FAO) supported the research effort. Researchers from Australia, Germany and The Netherlands came to Jamaica for extended periods to work with local staff. About the time the FAO project ended, the UK Government, through its Overseas Development Administration (ODA), supplied a research team and an electron microscope. Researchers from the University of Florida also worked in Jamaica. Research institutions in the UK and USA were also involved in the effort. Notable co-operation was received from the following organisations: IRHO (now CIRAD-CP), University of the West Indies, Unilever and IFA in Columbia.

External funding ceased in the early-1980s and culminated in considerable valuable information being obtained. Research on the nature of LY was done largely by visiting scientists while local staff concentrated mainly on plant improvement and management.

After the 1961 LY outbreak, a disease monitoring program was established in and around disease-areas. It was important to be certain which plants were dying from

LY. Monitoring also provided information on varietal resistance and data for replanting programs.

Felling of affected palms to check the spread of the disease was ineffective, as it had been earlier in the west, and was abandoned. Pesticides were used in an attempt to control new outbreaks before they could spread, but without success.

Once MLO had been found in diseased palms and it was realised that tetracycline could suppress LY symptoms and keep palms alive for years, chemotherapy was tried but rejected because of high costs and perceived health risks.

In 1961, the overwhelming majority of palms grown commercially were the susceptible 'Jamaica Tall' ecotype. The other two ecotypes being grown commercially, the 'Malayan Dwarf' and 'Panama Tall' had been introduced following hurricanes earlier in the century.

As has been indicated earlier, the 'Malayan Dwarf' was found to possess good resistance to LY. This ecotype, in addition to its resistance, is precocious and yields well under good conditions. Nevertheless, it has certain undesirable features: relatively low oil content and low yield stability. Therefore the search continued for other resistant germplasm. With assistance from ODA and FAO, germplasm collections were made from Southeast Asia and the Pacific regions and then screened for resistance to LY in field trials.

When it became obvious that none of the recent introductions were more resistant to LY than 'Malayan Dwarfs', a hybridisation program commenced to combine, in the F_1, the high disease resistance of the dwarf with the large fruit size and hardiness of the talls. Fortunately, one of the early crosses ('Malayan Dwarf' x 'Panama Tall') was productive and resistant, and a system was devised to produce it commercially.

By 1962, the resistance and productivity of 'Malayan Dwarf' were sufficiently proven that thousands of mother palms were selected for regular seed production. From that point, even in areas not yet attacked by the disease, planting of the 'Malayan Dwarf' began to supersede that of the susceptible 'Jamaica Tall'. By 1974, the 'Maypan' had shown sufficient disease resistance and productivity to be released to farmers.

In an attempt to speed up replanting, the Advisory Section of the Coconut Industry Board was expanded and seed-gardens and nurseries established at numerous sites to provide effective and efficient distribution of seedlings.

In 1966, the Coconut Industry Board initiated a Lethal Yellowing Rehabilitation Program financed by a levy on the crop. It provided selected resistant palms and cash instalments to growers losing tall palms to LY. Although growers had confidence in the advice given by the Board (because it was backed up by research), they still found it difficult to accept the idea that coconuts should have the regulation and attention of other crops. Nevertheless, by 1979, almost 3 million seedlings had been distributed under the program.

In a further attempt to increase production, a Copra Production Acceleration Program (COPRA) superseded the Lethal Yellowing Rehabilitation Program in 1980. Under this scheme, farmers obtained free seedlings and fertiliser, delivered to the farm gate, and a grant to help with weed control. When the program was terminated in 1989, about 10 million 'Malayan Dwarf' and 'Maypan' seedlings had been distributed.

4. Conclusions

Although the problem of LY has not been solved, the coconut industry of Jamaica has remained alive mainly by using planting material with good resistance. Nevertheless, an island-wide disease monitoring program is still in place.

LY still remains a threat. Since the hurricane on 1988 there have been outbreaks of the disease in pockets where losses of the 'Malayan Dwarf' and 'Maypan' have been much higher that previously observed.

There is an urgent need to find and utilise material other than the 'Malayan Dwarf' with good resistance so as to broaden the genetic base of the industry. Hand in hand with this, more effective methods must be found to control vector populations. It is hoped that the development of new techniques based on molecular biology, and advances in clonal propagation, embryo culture and cryopreservation will assist greatly in the war against lethal yellowing.

Acknowledgements

This paper was published with permission from the Coconut Industry Board, Jamaica.

References

1. Beakbane AB, Slater CHW and Posnette FA (1972) Mycoplasmas in the phloem of coconut, *Cocos nucifera L.*, with lethal yellowing disease. J Hort Sci 47:265
2. Been BO (1981) Observations on field resistance to lethal yellowing in coconut varieties and hybrids in Jamaica. Oléagineux 36:9-11
3. Frison EA, Putter CAJ and Dieckmann M, eds (1993) FAO/IBPGR technical guidelines for the safe movement of coconut germplasm. Rome: Food and Agriculture Organisation and International Board for Plant Genetic Resources
4. Howard FW, Norris RC and Thomas DL (1983) Evidence of transmission of palm lethal yellowing agents by a planthopper *Myndus crudus* (Homoptera: Cixiidea). Trop Agric 26:110-112
5. McCoy RE (1972) Remission of lethal yellowing in coconut palm treated with tetracycline antibiotics. Plant Dis Reptr 56:1019-1021
6. McCoy RE, Howard FW, Tsai JH, Donselman HM, Thomas DL, Bsham HG, Atilano RA, Eskafi FM, Britt L and Collins ME (1983) Lethal yellowing of palms. Univ Florida Agric Expt Sta Bull No 834
7. Nutman FJ and Roberts FM (1955) Lethal yellowing: unknown disease in Jamaica. Emp J Exp Agric 23:257-267
8. Plavsic-Banjac B, Hunt P and Maramorosch K (1972) Mycoplasma-like bodies associated with lethal yellowing disease of coconut palms. Phytopath 62:298-299
9. Romney DH (1971) The Malayan Dwarf supersedes the Jamaica Tall: costs and returns. Oléagineux 26:25-32
10. Romney DH (1971) Attempts to control lethal yellowing. 11th Ann Rep Res Dept Coconut Industry Board
11. Romney DH (1983) Brief review of coconut lethal yellowing. Indian Coconut J 13:1-8

Reproductive biology of coconut palms

G.R. ASHBURNER
Agriculture Victoria, Institute for Horticultural Development, Private Bag 15, South Eastern Mail Centre Vic. 3176, Australia
Present address: Centro de Investigación Científica de Yucatán, Apartado Postal 87, Cordemex, 97310 Mérida Yuc., México

1. Introduction

In order to develop appropriate management strategies for both *in situ* and *ex situ* conservation, studies of interactions between natural population structure, floral biology and pollination and seed dispersal vectors are needed [40]. Knowledge of these systems in coconuts is also relevant in the effective establishment of seed-gardens for hybrid seed production and for efficient crossing during a breeding program. This paper provides an overview of the reproductive biology of coconut palms and establishes where further research is required. A detailed review of the development of floral organs and fruit of coconut palms has been published by Menon and Pandalai [26].

2. Inflorescence structure

Coconut palms are monoecious and produce axillary, interfoliar inflorescences. The inflorescence consists of a central axis, the rachis, from which the rachillae (spikelets) arise in a spiral succession. At their apices, the rachillae bear staminate flowers in the axils of tertiary bracts, and bear pistillate flowers at their bases [22]. The number of female flowers in an inflorescence is dependent on both genetic and environmental factors [5] and varies from zero (with the first inflorescence produced by a palm) through to approximately 300 [26]. Similarly, the range of male flowers is from several thousand through to very few for some spicata inflorescences [26].

C. Oropeza et al. (eds.), Lethal Yellowing: Research and Practical Aspects, 111–121.

Each male flower has six perianth parts arranged in two whorls, the inner three alternating with the outer three. Enclosed in this floral envelope are six stamens which yield large quantities of powdery pollen. In the centre of each flower, there is a rudimentary pistil which divides into three segments at its apex, each bearing a nectar gland [26].

The female flower also consists of six perianth parts which are tightly folded over the bulbous pistil. Usually there is a male flower on each side of the female. A ring of six suppressed stamens is found in between the perianths and the fleshy ovary. The tip of the pistil has a raised nipple from which the stigma is exposed when the three sections separate at receptivity [26].

3. Phenology

3.1 Whole plant phenology

Mature coconut fruit take from 30 to 220 days to germinate, depending on both genotype [50] and environmental conditions [9]. Some dormancy mechanism must be involved, although it has not been fully characterised. The duration of juvenility is primarily influenced by genotype. This juvenile phase ranges from one year in very rare cases [31] through to 5-7 years for most tall ecotypes [39]. Maturity technically occurs when the first floral primordium is initiated, but is first observed by the emergence of the first inflorescence. In its natural state, generation time depends upon the seedling recruitment, but is largely irrelevant in an agricultural system. The degree of seedling recruitment is influenced by disturbing events such as cyclones or volcanic eruptions [33], or may simply be due to opening of the canopy during ageing [10].

3.2 Floral phenology

Development of the inflorescence has been summarised by Coomans [5] and studied in detail by Juliano [22] and Patel [32]. Floral initiation occurs approximately 33 months prior to inflorescence emergence, which is 4 months after the leaf primordium first differentiates from the apical meristem. The external spathe differentiates approximately 23 months before opening and the internal spathe differentiates 17 months before opening. The inflorescence becomes visible approximately 8 months before opening and usually opens in the axil of the 8th youngest open frond.

Development of the male and female flowers has previously been described [22, 23, 34] and will not be discussed in this paper. Male flowers start to mature as soon as the inflorescence opens, with the female flowers usually maturing after this. The relative timings of the male and female reproductive phases are discussed later in this paper.

Staminate flowers are mature when their bracts open and pollen is released from the anthers. Opening of the bracts usually occurs early in the morning with the flowers shedding pollen throughout the morning before dropping off after midday [1]. Staminate flowers tend to mature earlier at the top of the inflorescence and the top of each rachilla [36]. Coconut pollen is roughly spherical in shape when shed, but become ellipsoidal after contact with the atmosphere. In this condition, they are 0.065-0.069 mm x 0.028-0.030 mm in dimension [25]. Each anther contains between 111,000 and 221,000 pollen grains [25] giving approximately 272 million pollen grains in an inflorescence [1]. The pollen has been estimated to remain viable for between 2 days [25] and 8 days [28], but this requires further investigation.

Pistillate flowers become receptive early in the morning as is indicated by presentation of the stigmatic surface. Apart from stigma presentation, nectar containing 9-12% sucrose is produced from receptive flowers and continues to be exuded throughout the day [20]. The stigmas remain receptive to pollen for 1-3 days before degeneration. A complete study of coconut floral fertilisation has not been performed and this subject requires further examination.

3.3 *Fruit phenology*

Coconut fruit are technically described as fibrous drupes. The fruit take approximately 11-12 months to mature. The mesocarp and exocarp develops soon after fertilisation, followed by endocarp development after 4 months and endosperm development after 5 months [21]. The cytology of the fruit and fruit development has been well studied [6, 22, 43, 53]. Embryo development has been studied by Haccius and Philip [13] and the embryo growth rate appears to be directly related to endosperm deposition (Ashburner unpubl.).

4. Breeding System

4.1 Male and female phases

In a coconut inflorescence, the male phase (indicated by anthesis of staminate flowers) usually begins before the female phase (indicated by receptivity of pistillate flowers). However, the timing and duration of the female phase varies from occurrence during the male phase to occurrence post-anthesis (Fig. 1). An overlap of male and female phases between subsequent inflorescences is common. Rognon [35] classified the different flowering patterns found among coconut palms. In Type I flowering pattern, typified by some West African Tall ecotypes, no overlap of phases occurs (allogamy). Type II flowering pattern has no phase overlap in the same inflorescence but phase overlap between subsequent inflorescences, and is typified by Pacific Ocean Talls (mixed allogamy and indirect autogamy). In Type III flowering patterns, there is complete overlap of the phases within an inflorescence and is typified by the Malayan Dwarf complex (direct autogamy). Type IV flowering is characterised by partial overlap of the female phase within and between inflorescences (mixed). Within flowering groups, the degree of phase overlap is highly dependent on environmental conditions [38].

4.2 Self-compatibility

No self-incompatibility or partial self-incompatibility mechanisms have been observed in coconut palms [37, 42]. In fact, some dwarf ecotypes have shown an affinity for dwarf-type pollen [27, 37]. This observation may be due to increased pollen vigour as a consequence of selection for self-fertilisation. This area needs further investigation as it has important implications for seed-garden management.

4.3 Outcrossing rate

Using the fruit colour inheritance model postulated by Bourdeix [3], the maximum estimates of inbreeding in the Malayan Yellow Dwarf ecotype varies from 100% [19] through to 88.3% [51]. Few outcrossing estimates have been made for tall ecotypes but they are assumed to be preferentially outcrossing. This may be true for Type I flowering patterns, but in Type II, which is typical of Pacific Ocean

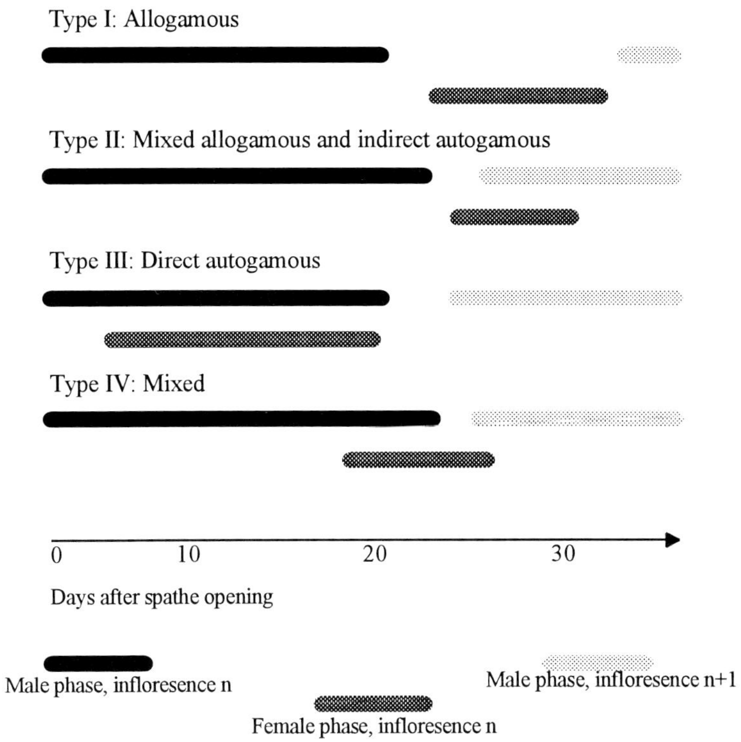

Fig. 1. Classsification of flowering patterns among coconuts [44].

types [24, 37, 42], self-fertilisation occurs at a relatively high rate [2], so a mixed mating system is probably involved.

5. Pollination mechanisms

With the exception of coconut palms, the sub-tribe Butiinae, which contains 9 genera [48], is exclusively beetle-pollinated [17]. The inflorescence of coconut has features that are typical of bee-pollinated palms. Such features are an open and laxly branching inflorescence consisting of sweet-scented protandrous flowers which become active during the day. The flowers are either pistillate or staminate, develop after bract-fall, open at dawn and drop around midday [17, 44]. Field observations support these generalisations. Although the inflorescences are visited

Table 1. Putative pollinators of coconut palms

Species	Order	Locality	Reference
Apis dorsata	Hymenoptera	Singapore	4
		India	47
Apis florea	Hymenoptera	India	47
Apis indica	Hymenoptera	Singapore	4
		Philippines	1
		India	47
		Indonesia	29
Apis mellifera	Hymenoptera	Hawaii	41
		Kenya	20
		Jamaica	11
		India	47
		Costa Rica	16
		Ecuador	16
Calliphora erythrocephae	Hymenoptera	Indonesia	29
Camponotus compressus	Hymenoptera	India	32
Homalictus cassiaefloris	Hymenoptera	Papua New Guinea	2
Homalictus dampieri	Hymenoptera	Papua New Guinea	2
Homalictus latitarsis	Hymenoptera	Papua New Guinea	2
India discolor	Lepidoptera	Indonesia	29
Lucilia sp.	Diptera	Philippines	1
Melipona sp.	Hymenoptera	Singapore	4
		Malaysia	7
Melipona fasciata	Hymenoptera	Costa Rica	16
Musca domestica	Diptera	Philippines	1
Polistes crinata	Hymenoptera	Jamaica	11
Polistes exclamans	Hymenoptera	Hawaii	41
Polistes macaensis	Hymenoptera	Hawaii	41
Polistes olivaceus	Hymenoptera	Hawaii	41
Rhynchiumatrum sp.	Coleoptera	Philippines	1
Trigona biroi	Hymenoptera	Philippines	1
Trigona corvina	Hymenoptera	Costa Rica	16
Trigona frontalis	Hymenoptera	Costa Rica	16
Trigona fulviventris	Hymenoptera	Costa Rica	16
Trigona tataira	Hymenoptera	Costa Rica	16
Vespaluctuosa sp.	Hymenoptera	Philippines	1

by a large variety of insects [41], bees are the major pollination vector with wasps playing a minor role (Table 1). Patel [32] also reported pollination through an ant vector. Coconut palms may also be pollinated by wind to a minor extent within a palm, but this is unlikely between palms [1,42]. Bat pollination of coconut palms [45] is likely to be minor since anthesis is usually reached during the morning [1].

Behaviour of the insects associated with the inflorescence is also important. Notable features are that there is competition for female flowers among pollinators and between pollinators and non-pollinators [11, 16, 42]. In Jamaica, Jay [20] determined that honey-bees collected either only pollen or only nectar, and that pollination probably occurred either due to pollen-foraging bees crossing the stigma, or due to transfer of pollen by nectar-foraging bees visiting both male and female flowers. Honey-bees cannot land on the slippery surface of female flowers, therefore they must hover to remove the nectar and, in so doing, brush up against the stigma. In Indonesia, Tammes [46] observed that the pollinating insects frequently returned to the same or a neighbouring palm the next day.

To prevent pollination from outside a seed-garden in Côte d'Ivorie, a 300 m exclusion zone is recommended [27], but in other areas, it probably depends on the pollinators present. The authors claimed that pollen was mainly wind-borne in their experiment, which appears contradictory to other researchers. This point illustrates the need to characterise the main pollinating vectors present in specific growing regions in order to establish correct seed-garden protocols.

Various protocols have been established for artificial pollination for breeding purposes [14, 28], and will not be discussed in this paper.

6. Dispersal mechanisms

Coconut palms are able to take advantage of a water dispersal mechanism. The fruit displays many adaptations to this role [15], including the ability to float in sea water for up to 110 days and still remain viable [8]. Evidence to support short-distance water dispersal is provided by the establishment of coconuts on the new volcanic island of Krakatau [18] and on the nuclear testing site at Bikini Atoll [52]. The presence of pre-human coconut pollen records in the Cook Islands [30] indicates that long distance dissemination of coconut fruit by floating is also likely. More recently, humans have acted as the main dispersal vector which is reflected by changes in the fruit characters of domesticated coconuts [15].

7. Fecundity

The number of viable fruit produced by a coconut palm depends upon the number of female flowers, the number of fruit surviving to maturity and the viability of those fruit. As previously mentioned, the number of female flowers depends on both genetic and environmental influences. Drought and nutritional stresses reduce the number of female flowers and may even cause inflorescence abortion. Furthermore, fruit may abort after fertilisation because of pests, diseases, water or nutritional deficiencies, or weather conditions [26]. Approximately 3% of all mature fruit are barren [12]. Genotype also influences the production of viable fruit, with the viabilities ranging from zero (in the case of makapuno type) through to 100%.

8. Conclusions

The reproductive biology of coconut palms has been well characterised at the cellular level, apart from the cytology and histology of the fertilisation event. The effects of weather conditions and cultural practices on reproductive capacity have already been defined, and agronomists and crop physiologists have the responsibility to extend this knowledge into predictive models for yield. Debate still exists over the breeding system of coconut palms. Clearly, there is variation between ecotypes and therefore reproductive biologists need to use molecular marker technology to clarify this situation for the whole range of ecotypes. Let us now proceed on the assumption that coconut palms have a mixed mating system. Further study should be given to the pollination vectors present in various locations, their behavioural characteristics and the male and female phases of the forms that are being routinely used. Such studies will aid in optimising management protocols for seed-gardens and conservation blocks. Controversy still surrounds the predominant dispersal mechanism for coconut fruit [49]. This area is far from being solely an academic study, since it may help to explain the genetic structure of coconut populations and help to optimise collecting and crossing strategies. This area of the reproductive biology of coconut palms may never be resolved, but it safe to say that early in the coconut time-scale, water dissemination predominated. Later on that time-scale, humans directed the development and dissemination of coconut forms which they considered useful.

Acknowledgements

This work was supported by the Australian Centre for International Agricultural Research in collaboration with the Cocoa and Coconut Research Institute, Papua New Guinea. Travel assistance was provided by Centro de Investigación Científica de Yucatán, México.

References

1. Albada VC (1921) The pollination of coconut. Philipp Agric 10:195-207
2. Ashburner GR (1994) Characterisation, collection and conservation of *Cocos nucifera* L. in the south Pacific. PhD thesis, University of Melbourne
3. Bourdeix R (1988) Etude du déterminisme génétique de la couleur du germe chez le cocotier nain. Oléagineux 43:371-374
4. Burkill IH (1919) Some notes on the pollination of flowers in the Botanic Gardens, Singapore, and in other parts of the Malay Peninsula. Gardens' Bull SS 2:165-170
5. Coomans P (1975) Influence des facteurs climatiques sur les fluctuations saisonnières et annuelles de la production du cocotier. Oléagineux 30:153-159
6. Cutter VM Jr, Wilson KS and Dubé GR (1952) The isolation of living nuclei from the endosperm of *Cocos nucifera*. Science 115:58-59
7. Dammerman KW (1929) The Agricultural Zoology of the Malay Archipelago. Amsterdam
8. Edmondson CH (1941) Viability of coconut seeds after floating in sea. Occ Pap B P Bishop Museum Hawaii 16:293-304
9. Foale MA (1993) The effect of exposing the germpore on germination of coconut. In: MK Nair, HH Khan, P Gopalasundaram and EVV Bhaskara Rao, eds. Advances in Coconut Research and Development, pp 247-252. New Dehli: Oxford and IBH
10. Foale MA, Ashburner GR and Friend D (1994) Canopy development and light interception of coconut. In: Foale MA and Lynch PW, eds. Coconut Improvement in the South Pacific, ACIAR Proceedings Series No 53, pp 71-73. Canberra: Australian Centre for International Agricultural Research
11. Free JB, Raw A and Williams IH (1975) Pollination of coconut (*Cocos nucifera* L.) in Jamaica by honeybees and wasps. Appl Anim Ethol 1:213-223
12. Gangolly SR and Nambiar MC (1953) Studies on the occurrence of barren nuts in the coconut. Indian Coconut J 7:5-19
13. Haccius B and Philip VJ (1979) Embryo development in *Cocos nucifera* L.: a critical contribution to a general understanding of palm embryogenesis. Plant Syst Evol 132:91-106
14. Harries HC (1976) Coconut hybridisation by the Policaps and Mascopol systems. Principes 20:136-147
15. Harries HC (1978) The evolution, dissemination and classification of *Cocos nucifera* L.. Bot Rev 44:265-319
16. Hedström I (1986) Pollen carriers of *Cocos nucifera* L. (Palmae) in Costa Rica and Ecuador (neotropical region). Revista Biol Trop 34:297-301
17. Henderson A (1986) A review of pollination studies in the Palmae. Bot Rev 52:221-259
18. Hill AW and Docters van Leeuwen W (1929) Germinating coconuts on a new volcanic island, Krakatoa. Nature 132:674
19. Jack HW and Sands WN (1929) Observations of the dwarf coconut palm in Malaya. Malay Agric J 17:140-170

20. Jay SC (1974) Nectar and honey collection by honeybees from coconut flowers. Bee World 55:105-111
21. Jayasuriya VUD and Perera RKIS (1985) Growth, development and dry matter accumulation in the fruit of *Cocos nucifera* L. var *nana* form *pumila*. Cocos 3:16-21
22. Juliano JB (1926) Origin, development and nature of the stony layer of the coconut (*Cocos nucifera*). Philipp J Sci 30:187-200
23. Juliano JB and Quisumbing E (1931) Morphology of the male flower of *Cocos nucifera* Linn.. Philip J Sci 45:449-458
24. Leach BJ (1977) Observations on flowering and morphology of coconuts in Western Samoa. Internal report. Rome: Food and Agriculture Organisation
25. Manthriratna MAPP (1965) Coconut pollen. Ceylon Coconut Quart 16:102-110
26. Menon KVP and Pandalai KM (1958) The coconut palm. Ernakalum: Indian Central Coconut Committee
27. Nucé de Lamothe M de, and Rognon F (1975) Pollinisation assistée et contamination par des pollens indésirables. Oléagineux 30:359-364
28. Nucé de Lamothe M de, Wuidart W, Rognon F and Sangare A (1980) La fécondation artificielle du cocotier. Oléagineux 35:193-205
29. Pardede D, Randriani E and Marfuati A (1986) Penyerbukan bunga kelapa di Sub Balai Penelitian Kelapa Pakuwon. Pemberitaan Penelitian Tanaman Industri 12:26-35
30. Parkes A (1991) Impact of prehistoric Polynesian colonisation: pollen records for Central Polynesia. Proc 17th Pacific Sci Congr, Honolulu, Abstr
31. Pillai RV, Nair RB, Mathew C, Bavappa KVA and Ramadasan A (1973) Studies on photoperiodic responsive reaction in coconut. J Plantation Crops 1:89-92
32. Patel JS (1938) The Coconut- A Monograph. Madras: Government Press
33. Pursglove JW (1972) Tropical Crops. Monocotyledons. London: Longman
34. Quisumbing E and Juliano JB (1927) Development of ovule and embryo sac of *Cocos nucifera*. Bot Gaz 84:279-293
35. Rognon F (1976) Biologie florale du cocotier. Durée et succession des phases mâles et femelles chez divers types de cocotiers. Oléagineux 31:13-18
36. Rognon F and Nucé de Lamothe M de (1978) Récolte et conditionnement du pollen pour la pollinisation des champs semenciers de cocotiers. Oléagineux 33:17-23
37. Sangare A (1981) Compétition pollinique et légitimité des semences produites dans les champs semenciers de cocotiers. Oléagineux 36:423-427
38. Sangare A, Rognon F and Nucé de Lamothe M de (1978) Les phases mâles et femelles de l'inflorescence de cocotier. Influence sur le mode de reproduction. Oléagineux 33:609-617
39. Santos GA and Sangare A (1992) Evaluation and utilization of coconut genetic resources. In: Coconut Genetic Resources. Papers of an IBPGR Workshop, Cipanas, Indonesia, 8-11 October, 1991. International Crop Network Series No. 8, pp 70-78. Rome: International Board for Plant Genetic Resources
40. Sedgley M and Griffin AR (1989) Sexual Reproduction of Tree Crops. London: Academic Press
41. Sholdt LL (1966) Insects associated with the flowers of the coconut palm, *Cocos nucifera* L., in Hawaii. Proc Hawaii Entomol Soc 19:293-296
42. Sholdt LL and Mitchell WA (1967) The pollination of *Cocos nucifera* L. in Hawaii. Trop Agric 44:133-142
43. Siegerman N (1951) Three-dimensional cell shape in coconut endosperm. Am J Bot 38:811-823
44. Silberbauer-Gottsberger I (1990) Pollination and evolution in palms. Phyton 30:213-233
45. Start AN and Marshall AG (1976) Nectarivorous bats as pollinators of trees in West Malaysia. In: J Burley and BT Styles, eds. Tropical Trees: Variation, Breeding and Conservation. London: Academic Press
46. Tammes PLM (1937) Over den bloei en de bestuiving van den klapper. Landbouw 13:74-89
47. Thampan PK (1982) Handbook on the Coconut Palm. New Dehli: Oxford and IBH

48. Uhl NW and Dransfield J (1987) Genera Palmarum. Lawrence: Bailey Hortorium and International Palm Society
49. Ward RG and Brookfield M (1992) The dispersal of the coconut: did it float or was it carried to Panama? J Biogeog 19:467-480
50. Whitehead RA (1965) Speed of germination, a characteristic of possible taxonomic significance in *Cocos nucifera* Linn.. Trop Agric 42:369-372
51. Whitehead RA, Thompson BE and Williams LV (1966) A genetic marker of use in coconut seed production. Oléagineux 21:15-16
52. Williams H (1967) Bikini nine years later. Sci J 3:38-53
53. Winton AL (1901) Anatomy of the fruit of *Cocos nucifera*. Am J Sci 12:265-280

Coconut variation and genetic resources

DANIEL ZIZUMBO and JOSÉ ARELLANO
Centro de Investigación Científica de Yucatán, Apartado Postal 87, Cordemex, 97310 Mérida Yuc., México

1. Introduction

Planting genetically resistant material is recognised as the best control method to control lethal yellowing of coconut palms (LY) (Been, this volume). Due to the rapid and devastating spread of LY-type diseases through different African and American coconut-growing regions, this paper focuses on possible ways and limitations on acquiring and maintaining adequate genetic collections.

Considering the available knowledge on coconut genetic diversity world-wide, this paper also analyses the evolutive factors which have contributed to coconut diversification and the patterns of variation that have resulted. Finally, the extent of the total variability which is contained in existing collections is evaluated, focusing on germplasm which is being challenged by LY in experimental trials.

2. Origin and distribution of the coconut palm

The tribe Cocoeae, which includes the coconut palm (*Cocos nucifera* L.), is predominantly distributed in South America. Only two of the nine genera in the subtribe Butiinae are found in other areas: *Jubaeopsis* in South Africa, and *Cocos* in the tropical zone [45]. Recognition of these facts, along with the discovery of fossil remains of *Cocos zeylanica* in New Zealand, led to the hypothesis that *Cocos* originated on the western coast of tropical American [8] and eventually spread by ocean currents to the Pacific Islands.

Uhl and Dransfield [45] proposed that the origin of the tribe Cocoeae was in western Gondwana. The tribe probably differentiated immediately prior to the fragmentation of that super-continent. Some of its members diversified in

C. Oropeza et al. (eds.), Lethal Yellowing: Research and Practical Aspects, 123–138.

America, others radiated to the African and Madagascan shelf where they still survive, and still others radiated to the Indian shelf, but are now extinct. The coconut fruit is capable of migrating in oceans due to its capacity for flotation, and therefore its dispersal became independent of continental drift. The discovery of the fossil remains of *Cocos sahnii* and the related species, *Palmoxylon sundaram*, in northern India, near the Himalayan Ranges, as well as the occurrence of the closely related *Voanioala gerardii* in Madagascar, possibly indicates that the original distribution of *Cocos* comprised the coastline and islands of the ancient Tethys Sea. Natural dissemination is thought to have taken place on emerging islands and the coastlines of the Indian and the Pacific Oceans [20].

The original distribution of coconut is thought to have included the Seychelles Isands to the west, the Line Islands to the east, the islands of Melanesia to the south and the Philippine Islands. to the north [14, 16].

Dissemination by floating probably resulted in the presence of small populations in remote places, in such a way that they became geographically isolated. It has been hypothesised that due to the founder effect and genetic drift, the limited gene flow allowed the fixation of a large number of genetic traits, leading to low intrapopulation diversity, high interpopulation diversity and to a wide range of adaptations to environmental factors [2, 10]. However, the extent of genetic migration in these putative populations is unknown. Natural selective pressures favoured morphological traits which increased fruit bouancy and rooting capacity in habitats with frequent movement and flooding. Long angular fruits with a high mesocarp content and seeds with a slow germination rate were probably selected [16].

3. Domestication and artificial dissemination

Harries [19] believed that the Malaysian floristic region [13] was where humans first interacted with the coconut palm, thus initiating the domestication process. Coconuts are thought to have been growing on coral atolls and would have provided a simply obtained source of potable water, and played an important role in human migration in this area. Additional evidence to support this hypothesis is the discovery of fossil coconut shell and root fragments (5 420 B.P.) in Vanuatu [43], the observation of wild coconuts in northern Borneo [18], and the association of coconut remains, with human settlements in the San Matias Islands, Papua New Guinea and in the Solomon Islands (4 500-2 500 B.P.) [24].

Man became a major evolutive force by deliberate dissemination, cultivation and selection of coconut planting material. Selection pressures under domestication resulted in morphological traits such as fruits with larger seeds (low mesocarp percentage) and seeds with a high endosperm percentage (especially liquid endosperm). A reduction in mesocarp percentage from 70% to 35% result, This leading to a short, less angular fruit with an almost round shape. Fruit exhibiting this morphology are unable to float well and are therefore disadvantageous for natural dispersal. These fruit contain a nearly spherical or oval seed [16]. Selection of fruit with high water contents resulted in fruit with an endosperm proportion of up to 50%, but this modification was not at the expense of endocarp proportion. Based on this array of desirable morphological traits, visual selection could have easily been applied in a conscious way to every generation of coconut palms [19].

It is probable that selected coconut palms were growing near human settlements close to the beach, or were cultivated inland along river watercourses, and remained isolated from wild populations. Selection of planting material may have favoured seednuts with a low mesocarp percentage because they germinate faster and can be placed in an upright position in the seedbed, further increasing the speed of germination. In addition, seedling selection below mother palms could also favour early germination, as those individuals tend to overshadow those seednuts that sprout later giving rise to seedlings that die or remain small [17, 19].

Man overcame previous natural barriers to coconut migration and disseminated the coconut palm world-wide thereby promoting different ecotypes to grow in close association. This, in turn, led to an increase in intrapopulation diversity through hybridisation and introgression [16]. Malay and Arab voyagers are thought to have carried coconuts to Southeast Asia, India and East Africa approximately 2 000 years ago [42]. Similarly, Polynesians are thought to have distributed domesticated coconuts in the South Pacific Islands (1 000 B.P.) and the Portuguese and Spanish to West Africa and America (500 B.P.) [49]. During the 1800s, the Germans and British further disseminated the species among their colonies, and during the past 60 years, many coconut-producing countries have disseminated it within their respective territories [16]. Over the last few decades, extensive cultivation of high yielding ecotypes and hybrids has occurred at the expense of local landraces and wild-types.

The variability of coconut palms has been interpreted by Harries [16] as a product of introgressive hybridisation between coconut populations showing "wild" and "domesticated" characters, which then gave rise to a *continuum* of morphologies that ranged between the two contrasting populations. The diversity found on a world-wide basis suggests that, after domestication, several secondary centres of diversity arose [29, 30]. These secondary centres include East Africa and India, where coconuts with "wild" characters have been found; the Indo-Malaysian region, where coconuts with "domesticated" characters are widespread; and Melanesia and Polynesia, where both types have been described. For the last region, Ashburner [2] found that the genetic pool is actually formed by populations resembling both types (wild and domesticated). Using morphological and molecular techniques, he identified the original gene pool, composed by wild type populations, broadly distributed from Papua New Guinea through the south Pacific Islands. The domesticated gene pool overlays the wild gene pool and is characterised in intensity by a clinal distribution with the degree of domestication decreasing across the Pacific to the east. However, relatively pure domestic populations also occur in Rennell Island, and the Marquesas and Hawaii Isands.

4. Development of germplasm collections

Over the last century, there has been much germplasm assemblage and maintenance. Initially, coconut germplasm collections were directly related to breeding programs conducted in agricultural experimental stations in Indonesia and Madagascar, from 1901 to 1910. World War I ended those activities in both countries. Breeding schemes started with the production of hybrids in Fiji (1928), India (1932), New Guinea and in Sri Lanka (1938), but were abandoned due to the great depression and World War II [34].

Presumably, the amount of genetic variability contained in those collections was low, and only represented a few accessions which showed desirable agronomic traits. The primary objectives of the breeding programs were to increase copra yields, therefore exploration and collection activities were oriented towards the selection of healthy individual palms with high seednut production. Samples were taken from different bunches on the same palm and then maintained as long-lived plantations in agricultural stations. Unattended plantations, wild populations and individual palms showing undesirable characteristics were usually ignored.

After World Ward II, coconut germplasm collections registered a noticeable growth in different regions. Presently, due to maintenance requirements, a sufficient amount of gene banks are needed in different countries for breeders to draw on, depending on particular breeding aims: high yields, product quality, disease tolerance or adaptation to particular habitats. Improvement of breeding and conservation strategies have influenced the activities of the germplasm collections, and there is now a growing concern to adequately sample the variability of the species. Exploration and the methodical collection of wide areas has been accomplished, although little attention has been given to wild populations.

At the present time, several germplasm collections have been established in Africa in the Ivory Coast, Ghana, Nigeria and Tanzania (Table 1). The foremost collection is located in the Ivory Coast, where the Marc Delorme Station maintains a collection which is comprised of a total of 61 populations. These populations are derived from the Pacific Islands, America, South Asia, Southeast Asia and West Africa [31, 39]. Nigeria has concentrated on indigenous germplasm, with 12 local 'West African Tall' (WAT) populations included in the collection [1]. The Ghanan and Tanzanian collections are oriented towards selecting germplasm resistant to the lethal diseases caused by MLOs. Ghana is currently testing 11 populations and Tanzania 21 populations, many of them were introduced from the Marc Delorme Station [40, 41].

In Southeast Asia, three important collections can be identified in the following countries: Thailand, Indonesia and the Philippines (Table 1). These germplasm collections contain a high number of local ecotypes from South and Southern Asia. There is no material from America and very few accessions have been introduced from Africa, and those simply represent those ecotypes relevant to breeding programs, such as WAT and 'Cameroon Red Dwarf' (CRD) [27, 38, 44].

In southern Asia, India and Sri Lanka have also assembled germplasm collections (Table 1). In India, the Central Plantations Crops Research Institute (CPCRI) established a coconut germplasm collection in two locations: Kasaragod and Sipighat. The collection is comprised of 126 populations, 40 indigenous and 86 exotic from 22 countries, with emphasis on Southeast Asia and the Pacific Islands [26, 36, 37].

In the Pacific region, there are important germplasm collections in Vanuatu, Papua New Guinea and the Solomon Islands (Table 1). Indigenous populations are represented in these collections, as well as some ecotypes from Southeast Asia. Representative material from Africa and America was introduced from the Ivory Coast [6, 11, 32]. Fiji, American Samoa, Cook Islands, French Polynesia, Kiribati, Tonga, Tuvalu and Western Samoa, have established small breeding collections on the basis of imported germplasm and the following populations have been

Table 1. Coconut germplasm collections throughout the world.

Region/ country	Collection	Variety	No pops	No palms	No palms/ pop	< 40 inv/ pop	No populations from the region							Reference
							East Africa	West Africa	South Asia	Southeast Asia	Pacific	East America	West America	
AFRICA														
Ivory Coast	Marc Delorme	Tall	42	6,496	155	4	1	5	7	17	10	0	2	39
		Dwarf	19	3,605	189	1	0	1	1	11	3	1	0	
Tanzania	Zanzibar		13	1,197										35
	Mafia		13	1,194										35
	NCDP	Tall	14	4,113	196	0	3	1	0	4	5	0	1	41
		Dwarf	7	2,957	180	0	0	3	2	2	0	0	0	
Ghana	Ghana	Tall	6	52	9	6	0	1	2	3	0	0	0	40
		Dwarf	5	52	10	5	0	2	1	2	0	0	0	
Nigeria	Nigera	Tall	12	-	-	-	0	12	0	0	0	0	0	1
SOUTHEAST ASIA														
Thailand	Chumphon	Tall	10	4,535	454	1	0	1	1	7	1	0	0	44
		Dwarf	10	2,015	202	3	0	1	0	8	1	0	0	
Indonesia	Mapanget	Tall	30	1,883	63	4	0	1	0	27	2	0	0	27
		Dwarf	9	515	57	2	0	0	0	9	0	0	0	
	Tamnes	Tall	1	571	571	0	0	0	0	1	0	0	0	27
	Pakuwon	Tall	13	880	68	3	0	0	0	11	2	0	0	27
		Dwarf	10	429	54	2	0	0		10	0	0	0	
	Bone Bone	Tall	35	1,797	51	10	0	0	0	35	0	0	0	27
	Selakau	Tall	7	830	119	0	0	0	0	7	0	0	0	27
		Dwarf	1	110	110	0	0	0	0	1	0	0	0	
	Makariki	Tall	6	360	60	0	0	0	0	6	0	0	0	27
		Dwarf	1	60	60	0	0	0	0	1	0	0	0	

	Marihat	Tall	23	6,687	278	1	0	1	0	18	4	0	0	38
		Dwarf	12	4,050	162	1	1	1	0	10	0	0	0	
Philippines	Zamboanga	Tall	52	5,589	107	3	0	1	0	45	6	0	0	Santos
		Dwarf	25	4,050	162	2	0	1	1	22	1	0	0	pers com
SOUTH ASIA														
India	Kasaragod/	Tall	99	1,559	16	85	2	3	32	14	42	5	1	26
	Sipighat	Dwarf	27	606	22	21	0	1	1	22	1	0	0	
PACIFIC														
Vanuatu	Saraoutu	Tall	21	3,606	172	0	0	1	0	4	16	0	0	6
		Dwarf	13	1,770	136	1	0	1	0	7	4	1	0	
Papua New Guinea	CCRI	Tall					0	0	0	0	10	0	0	32, 33
		Dwarf					0	0	0	2	8	0	0	
Solomon Islands	Yandina	Tall	14	1,206	86	4	0	0	1	1	12	0	0	9, 11
		Dwarf	7	345	49	4	0	0	0	1	6	0	0	
AMERICA														
Jamaica	CIB	Tall	26	1,581	59	18	1	0	3	5	14	1	2	4
		Dwarf	10	1,230	123	3	0	0	4	3	2	1	0	
México	CICY	Tall	26	2,873	111	1	0	1	0	0	2	4	19	
		Dwarf	5	350	70	1	0	0	0	5	0	0	0	
Brazil	CNPCo	Tall	8	1,520	190	2	0	1	0	1	5	1	0	7
		Dwarf	6	1,964	327	0	0	1	0	5	0	0	0	

introduced: 'West African Tall', 'Rennell Tall', 'Polynesian Tall' and 'Malayan Yellow Dwarf' [2].

In America, there are germplasm collections in Jamaica, Brazil and México (Table 1). The foremost collection was established in Jamaica, between 1962 and 1965, to test the response of different genotypes to lethal yellowing. A total of 36 populations were collected in the Pacific Islands, South and Southeast Asia, Africa and America [3, 46, 48]. During the 1980s, Brazil initiated a breeding collection which includes one indigenous population and seven parental populations imported from the Ivory Coast [6]. México has a collection for the purpose of determining the resistance levels of indigenous germplasm against lethal yellowing. It currently comprises 23 populations collected on the Pacific and the Atlantic coasts of México, as well as some introductions from the Ivory Coast [51].

The present status of coconut germplasm collections in the world is summarised in Table 1. According to the origin of collected material, it can be seen that almost 45% of the populations are from Southeast Asia and 30% from the Pacific region. The remaining 25% are derived from South Asia, Africa and America. This situation indicates that more collection has occurred in areas with seemingly high diversity, resulting from the reputation of some of the ecotypes from Southeast Asia and the Pacific. In addition, some ecotypes derived from Southeast Asia and the Pacific have shown resistance to lethal yellowing (LY) (*e.g.* 'Malayan Red and Yellow Dwarf', 'Rotuma Tall', 'Cambodia Tall', 'Bougainville Tall' and 'Malayan Tall').

It is difficult to assess how well the current germplasm repositories represent the total diversity of the species, as there is a general lack of knowledge of its genetic structure and variability [5]. A study from the Pacific region indicated that interpopulation diversity in that region has been undercollected, with only a limited number of the divergent populations actually represented in established collections [2]. It is likely that a similar situation occurs in South and Southeast Asia. African and American genetic diversity has been considered lower and therefore is also not adequately represented in collections. In summation, interpopulation diversity is not well represented and the low diversity contained in collections is restricted to cultivated populations exhibiting currently desirable traits.

Intrapopulation diversity in most collections consists of more than 100 palms per population (Table 1). Marshall and Brown [25] calculated that 95% of a population genetic variability in most open pollinated crops should be represented in a sample of 50 to 100 individuals selected at random. Using molecular characters, Ashburner [2] estimated that between 7 (for some dwarfs) and 37 individuals would give a good representation of the diversity of coconut palm populations in the Pacific region, depending on the populations sampled. Using these criteria, it appears that most of the collections have adequate intrapopulation diversities, except those cases where most of the accessions consist of a very low number of palms per population (less than 40 palms).

With regards to the genetic diversity which is being tested against LY-type diseases (Tables 2 and 3), it can be seen that 27 tall and 10 dwarf ecotypes, as well as 47 hybrid combinations are currently under evaluation in experimental resistance trials in Jamaica, Tanzania and Ghana. Some resistance has been found in the trials established in Jamaica, this includes the production of the 'Maypan' hybrid as a result of crossing the 'Malayan Dwarf' with 'Panama Tal'l. This hybrid is both highly resistant and productive and has been used on a commercial scale to replace local populations of the highly susceptible 'Jamaica Tall' palms thereby resurrecting the domestic coconut industry. Preliminary results derived from experiments conducted in Tanzania and Ghana can be used to contrast current knowledge on the sources of resistance (see also Harries, this volume). Observations made in Jamaica and Ghana coincide to indicate that the 'Malayan Yellow Dwarf', 'Malayan Red Dwarf' and 'Sri Lanka Green Dwarf' varieties are highly resistant to the LY-type diseases. Tall coconut populations from Southeast Asia and from the Pacific are less resistant. Wild-type populations such as the 'Jamaica Tall', 'West African Tall', 'Vanuatu Tall' and 'India Tall', including their hybrid progeny, are highly susceptible. Conversely, observations derived from experimental trials in Tanzania do not confirm this pattern. In Tanzania, the dwarf varieties are not highly resistant and their resistance levels are similar to some less resistant tall ecotypes. The most promising source of resistance has been identified in an indigenous wild-type population of 'East African Tall' (Table 2).

These differential results can be explained in a number of different ways. Discrepancies may arise by differences in the genetic composition of the causal agent of the disease in the different regions [22]. Also, they can be due to the existence of differences in the genetic composition of the coconut populations under evaluation, since most of them have different origins. The 'Malayan Dwarf' populations tested in Jamaica and Tanzania were derived from different and

Table 2. Resistance levels of coconut ecotypes to LY-type diseases [3, 40, 41]. HR = highly resistant, LR = less resitant, LS = less susceptible and HS = highly susceptible.

Form	Source	Ecotype	Trial location and resistance level		
			Jamaica	Tanzania	Ghana
Tall	East Africa	East African Tall (EAT)	-	LR	-
Tall		Seychelles Tall	LS	-	-
Tall	West Africa	West African Tall (WAT)	-	LS	LS
Tall	South Asia	Indian Tall	HS	-	-
Tall		Sri Lanka Tall	LS	-	-
Tall	Southeast Asia	Malayan Tall (MLT)	LR	LS	-
Tall		Thailand Tall	LR	LR	-
Tall		Cambodia Tall	LR	LR	-
Tall		Tagnanan Tall	-	LR	-
Tall	Pacific	Vanuatu Tall (VTT)	HS	LR	-
Tall		Rennell Tall (RLT)	LS	LS	-
Tall		Polynesian Tall (PYT)	LS	LS	-
Tall		Kar Kar Tall (KKT)	LR	LR	-
Tall		Rotuma Tall	LR	LR	-
Tall		Bougainville Tall	LR	-	-
Tall		Yap Tall	LR	-	-
Tall		Sarawak Tall	LR	-	-
Tall		Fiji Tall	LS	-	-
Tall		Markham Valley Tall	LR	-	-
Tall		Rangiroa Tall	LS	-	-
Tall		SolomonTall	LS	-	-
Tall		Samoa Tall	LS	-	-
Tall		Tonga Tall	LS	-	-
Tall	East America	Jamiaica Tall (JT)	HS	-	-
Tall	West America	Panama Tall	LR	LR	-
Tall		Peru Tall	LR	-	-
Dwarf	East Africa	Pemba Red Dwarf (PRD)	-	LR	-
Dwarf	West Africa	Cameroon Red Dwarf (CRD)	-	LR	LR
Dwarf		Equitorial Green Dwarf (EGD)	-	LS	HR
Dwarf	South Asia	Sri Lanka Green Dwarf (SGD)	HR	LR	HR
Dwarf		Indian Dwarf	HR	-	-
Dwarf	Southeast Asia	Malayan Red Dwarf (MRD)	HR	LR	LR
Dwarf		Malayan Yellow Dwarf (MYD)	HR	LR	HR
Dwarf	Pacific	Niu Leka	LR	-	-
Dwarf		Rangiroa Dwarf	HS	-	-
Semi-dwarf	South Asia	King coconut	HR	LR	-

Table 3. Resistance levels to MLO diseases of hybrids tested in different countries. HR = highly resistant, LR = less resitant, LS = less susceptible and HS = highly susceptible.

Hybrid	Trial location and resistance level			Hybrid	Trial location and resistance level		
	Jamaica	Tanzania	Ghana		Jamaica	Tanzania	Ghana
WAT X RLT	-	LR	-	EGD X VTT	-	LR	LR
MLT X MLT	-	-	LR	SGD X WAT	-	-	LR
VTT X MLT	-	-	LR	MRD X PYT	-	LR	LR
VTT X VTT	-	-	HR	MYD X EAT	-	LR	HR
RLT X WAT	-	-	LR	MYD X MZT	-	LS	-
RLT X RLT	-	-	LS	MYD X WAT	LR	-	-
PYT X PYT	-	-	LR	MYD X MLT	-	LS	LS
JT X MZT	LS	-	-	MYD X CAT	HR	-	-
JT X CT	LS	-	-	MYD X VTT	-	LR	HR
JT X RLT	HS	-	-	MYD X RLT	LR	-	LR
JT X PYT	LS	-	-	MYD X PYT	LR	LS	LR
JT X SLT	LS	-	-	MYD X SLT	LR	-	-
JT X PT	LR	-	-	MYD X JT	LR	-	-
PT X RLT	LR	-	-	MYD X PT	HR	-	-
PT X SLT	LR	-	-	NLD X JT	HR	-	-
PT X JT	LR	-	-	WAT X SGD	-	LS	-
CRD X EAT	-	LS	-	JT X CRD	HR	-	-
CRD X WAT	-	LS	LR	JT X MYD	LR	-	-
CRD X MLT	-	LS	LR	JT X CUD	LR	-	-
CRD X VTT	-	-	LR	PT X CRD	LR	-	-
CRD X RLT	-	LS	HR	PT X MYD	HR	-	-
CRD X PYT	-	-	LR	MYD X CRD	HR	-	-
EGD X EAT	-	LS	-	NLD X NLD	HR	-	-
EGD X WAT	-	LS	LR				

independent introductions from Malaysia [41, 47]. Recent studies using DNA characters have shown that different accessions of Malayan Yellow Dwarfs are distinct at the molecular level (Rohde pers. comm.). The local tall coconut populations found in Tanzania, Ghana and Jamaica display "wild" type characters, but they are not expected to be exactly the same genotypes, since their spread to these areas has been traced back to different periods and was accomplished through different processes [15, 42]. It has also been shown that different wild-type accessions are genetically distinct [2]. If possible, the existence of discrepancies should be assessed using available developments in LY diagnostic techniques (PCR analysis/DNA probes) and introducing some modifications in the experimental designs, to test materials from a common source in different sites. It is important to consider the need to increase the genetic diversity under evaluation to find alternative sources of resistance. In addition, it is necessary to use molecular or genetic markers which could be applied for variety identity (Ashburner, this volume).

5. Discussion

Morphological and molecular studies on the variation patterns of the "wild" gene pool indicate that it is presently distributed in a great number of small islands, atolls and shorelines corresponding to its natural distribution range, where contact with Man has been scarce [2]. Artificial selection, dissemination and cultivation of domesticated populations in areas where wild populations existed, led to a process of introgressive hybridisation, which, in turn, produced intrapopulation variability and the development of secondary centres of diversity [50]. Development of large commercial plantations at the end of the 19th century brought about selection for large-fruited and high-yielding populations to the detriment of wild populations, some of which are now in danger of disappearing [28]. Little interpopulation diversity of "wild" genotypes is contained in present collections, therefore there is a need to increase the number of accessions from islands and coastal areas in the natural distribution range.

Present collections do not adequately represent the diversity found in secondary centres, where ecotypes with different origins have hybridised, such as with the western coasts of tropical America and the Pacific Islands. Identification of areas such as these is necessary to prospect and to collect available diversity and to maintain it in germplasm collections. Although some work has been done on the effective management of coconut germplasm in a repository [2], further work is required.

Rapid spread of LY-type diseases in Africa and America has facilitated research to determine the relationship between the causal agents in each instance (Harrison *et al.*, this volume; Jones *et al.*, this volume). It is imperative to increase the genetic diversity under evaluation in resistance trials. Available results show that some resistance has been found in tall and dwarf ecotypes from the Malaysian floristic region. However, there are a lot of uncollected populations that require proper evaluation in the already established experimental trials and priorities must be placed on making them accessible, since there are currently difficulties due to phytosanitary restrictions [12] (Dollet, this volume) and limited financial support.

In both Africa and America, LY-type diseases pose a threat to the indigenous coconut populations with high productivity and agronomic performance. Appropriate methods must be developed to adequately preserve and utilise this germplasm before it is lost.

Strategies are required to strengthen research on biochemical, molecular and genetic markers, to assist genotype identification and recognition of resistant and high yielding material (see Ashburner, this volume). When these markers become available, they could be applied in genetic and progeny trials in improvement programs and to conduct collection of appropriate populations showing desirable traits.

The representation of genotypes from different geographic areas in established collections is very low. Germplasm exchange should be given high priority for both breeding and conservation purposes. Sources of resistance to LY and to other pests and diseases illustrate that it is becoming increasingly important for breeders to have access to all available germplasm for short-term programs. In the past, germplasm exchange has been limited due to the high costs for collection and transportation and to the technical aspects related to phytosanitary restrictions in different countries. Appropriate molecular methods are now available to estimate effective population size, amount of variation, phytosanitary certification for seednuts and pollen (indexing) and *in vitro* embryo culture techniques which can be applied to germplasm movement [2].

Establishing and maintaining a coconut germplasm collection is a complex, expensive and enduring task, mainly due to the long life-cycle and the large areas that must be devoted to experimental trials. Nevertheless, collaboration and co-ordination between existing collections is very important and could be improved through regional programs integrating strategies for *ex situ* and *in situ* conservation and germplasm use.

Ex situ collections have been used to evaluate adaptations to particular environmental conditions, to compare the performance of different ecotypes and are also a basis for germplasm use. They are also important because the materials already collected are being used to understand the behaviour of productivity, adaptive and resistance markers. However, the recurrent expenditure for the effective maintenance of a germplasm collection is high. Genetic variability can be reduced if regeneration is not properly conducted. Contamination, sampling and genetic drift are all potential risks to diversity if the long-term conservation is not carried out properly. Development of appropriate schemes for *in situ* conservation should solve some of the aforementioned problems. *In situ* conservation is less expensive, plants show adaptations to local environmental conditions and population size is large enough to prevent genetic erosion.

Implementation of both *in situ* and *ex situ* strategies could occur through the efficient exchange of information and germplasm. Monitoring and documenting each collection could be conducted through computerised records and with the use of adequate software, a database can be developed. This could be accomplished using the guidelines available from the International Plant Genetic Resources Institute [23], the International Coconut Cultivar Registration Authority [21] and the Coconut Genetic Resources Network [6].

Acknowledgements

This research was partially funded by CONACyT, through project 0598-N9109. Mr. Hugh Harries, M.C. Patricia Colunga, Dr. Carlos Oropeza and Dr. Roger Ashburner contributed to the revision of early versions of this paper and made important suggestions. The personal communication of Dr. Wolfgang Rohde of the Max Planck Institute is gratefully acknowledged.

References

1. Akpan EEJ 1994 Evaluation of tall coconut (*Cocos nucifera* L.) genotypes within the Nigerian coconut germplasm bank. Oléagineux 49:13-20
2. Ashburner GR 1994 Characterisation, collection and conservation of *Cocos nucifera* L. in the south Pacific. PhD thesis, University of Melbourne
3. Been BO 1981 Observations of field resistance to lethal yellowing in coconut varieties and hybrids in Jamaica. Oléagineux 36: 9-11
4. Been BO 1992 Status of existing coconut collection in Jamaica. In : CIRAD/IBPGR, eds. Report of a working session for implementation of the International Coconut Database. Montpellier: CIRAD/IBPGR
5. Bourdeuix R, N' Cho P, Le Saint J and Sangre A 1990 A coconut (*Cocos nucifera* L.) selection strategy I. Rundown of achievements. Oléagineux 45:366-371
6. CIRAD 1992 Report of a working session for implementation of the International Coconut Database. Montpellier: CIRAD/IBPGR

7. CNPCo 1992 Coconut germplasm collection of Brazil. In: CIRAD/IBPGR, eds. Report of a working session for implementation of the International Coconut Database. Montpellier: CIRAD/IBPGR
8. Cook OF 1910 History of the coconut in America. Contr US Nat Herb 14:271-342
9. Firman I, Jackson G and Guarino L 1986 Guidelines for the transfer of coconut germplasm and an inventory of South Pacific collections. RAS/83/001 Field Doc 10. Suva: Food and Agriculture Organisation
10. Foale MA 1992 Coconut genetic diversity present knowledge and future research needs. In: IBPGR, ed. Coconut Genetic Resources. Papers of an IBPGR workshop, Cipanas, Indonesia. 8-11 October, 1991. International Network Series No. 8, pp 46-53. Rome: IBPGR
11. Friend D 1975 The Yandina coconut collection. Yandina: Ministry of Agricultural and Rural Economy
12. Frison E.A. 1992. Safe movement of coconut germplasm: Presentation of guidelines. in: IBPGR. Coconut Genetic Resources. Papers of an IBPGR workshop, Cipanas, Indonesia 8-11 October, 1991. International Network Series No. 8. IBPGR, Rome. pp. 43-45.
13. Good R 1974 The geography of the flowering plants. Longman: Longman
14. Gruezo W S and Harries HC 1984 Self-sown wild-type coconuts in the Philippines. Biotropica 16:140-147
15. Harries HC 1977 The Cape Verde Region (1499-1549); the key to coconut culture in the Western Hemisphere? Turrialba 27:227-231
16. Harries HC 1978 The evolution, dissemination and classification of *Cocos nucifera* Linn. Bot Rev 44:265-319
17. Harries HC 1981 Germination and taxonomy of coconut. Ann Bot 48:873-883
18. Harries HC 1982 The antiquity of coconut palm in Western Borneo. Sarawak Mus J 24:239-242
19. Harries HC 1990 Malesian origin for a domestic *Cocos nucifera*. In: P. Baas *et al.,* eds. The Plant Diversity of Malesia, pp 351-357. Dordrecht: Kluwer
20. Harries HC 1992a Biogeography of coconut (*Cocos nucifera* L). Principes 36:155-161
21. Harries HC 1992b The ISHS coconut registration authority. In: IBPGR, ed. Coconut Genetic Resources. Papers of an IBPGR workshop, Cipanas, Indonesia. 8-11 October, 1991. International Network Series No. 8, pp 41-42. Rome: IBPGR
22. Harrison NA, Richardson PA, Jones P, Tymon AM, Eden-Green S and Mpunami AA 1994 Comparative investigation of mycoplasmalike organisms associated with Caribbean and African coconut lethal decline diseases by DNA hybridizations and PCR assays. Plant Dis 78:507-511
23. IBPGR 1992. Coconut Genetic Resources. Papers of an IBPGR workshop, Cipanas, Indonesia. 8-11 October, 1991. International Network Series No. 8. Rome: IBPGR
24. Kirch PC 1985 Lapita and oceanic cultural origins. Excavations in Mussau Island, Bismarck Archipelago. Seattle: University of Washington
25. Marshall DR and Brown AHD 1975 Optimum sampling strategies in genetic conservation. In: O.H. Frankel and J.G. Hawkes, eds. Crop Genetics Resources for Today and Tomorrow, pp 53-80. Cambridge: Cambridge University Press
26. Nair MK 1992 Coconut Genetic Resources. Coconut Res Dev 8:20-35
27. Novarianto HT Rompas and Darwis SN 1994 Coconut breeding programme in Indonesia. Coconut Research Institute. Manado: Central Research Institute for Industrial crops.
28. Nucé de Lamothe M de 1992 Coconut improvement: needs and opportunities. In: IBPGR, ed. Coconut Genetic Resources. Papers of an IBPGR workshop, Cipanas, Indonesia. 8-11 October, 1991. International Network Series No. 8, pp. 31-34. Rome: IBPGR
29. Nucé de Lamothe M de and Wuidart W 1979 Les cocotiers grands à Port-Bouët (Côte-d'Ivore) 1. Oléagineux 34:310-318
30. Nucé de Lamothe M de and Wuidart W 1981 Les cocotiers grands à Port-Bouët (Côte-d'Ivore) 2. Oléagineux 36:353-365

31. Nucé de Lamothe M de L, Sangare A, Meunier J and Le Saint JP 1991 Coconut hybrids. Interest and prospects; IRHO contribution to research and development. In: EG Silas *et al.*, eds. Coconut breeding and management, PP. 26-38. Trichur: Kerala Agricultural University
32. Ovasuru T 1992 Coconut germplasm collection in Papua New Guinea. In: CIRAD/IBPGR, eds. Report of a working session for implementation of the International Coconut Database. Montpellier: CIRAD/IBPGR
33. Ovasuru T 1994 The current status of the coconut industry in Papua New Guinea. In: MA Foale and PW Lynch, eds. Coconut Improvement for the South Pacific. ACIAR proceedings no 53, pp 9-13. Canberra: Australian Centre for International Agricultural Research
34. Ohler G J 1984 Coconut, tree of life. FAO Plant Production and Protection Paper 57. Rome: FAO
35. Persley GJ 1992 Replanting the tree of life. Oxford: CAB International
36. Pillai RV, Rao EVVB and Kumaran KB 1991 Characterisation of coconut cultivars. In: EG Silas *et al.*, eds. Coconut Breeding and Management, pp. 74-82. Trichur: Kerala Agricultural University
37. Rao MBS and Koshy PK 1981 Coconut germplasm collection in the Pacific Ocean Islands. FAO Consultant Report.
38. Samosir YMS 1993 Coconut collecting in Indonesia. Plant Genet Resour Newsl 91/92:58-59
39. Sangare A 1992 Current status of coconut genetic resources research in the Ivory Coast. In: CIRAD/IBPGR, eds. Report of a working session for implementation of the International Coconut Database. Montpellier: CIRAD/IBPGR
40. Sangare A, Taffin G de, Franqueville H de, Arkhust ED and Pomier M 1992 Coconut lethal yellowing in Ghana: initial results on planting material performance in the field. Oléagineux 47:699-704.
41. Schuiling M, Mpunami A, Kaiza D.A. and Harries HC 1992 Lethal disease of coconut palm in Tanzania III. Low resistance of imported germoplasm. Oléagineux 47 12:693-698.
42. Schuiling M and Harries HC 1994 The coconut palm in East Africa 1. East African Tall. Principes 38:4-11
43. Spriggs MJT 1984 Early coconut remains from the south Pacific. J Polynes Soc 93:71-77
44. Thirakul A, Petchpiroon C and Rattanaprut M 1992 Existing status of coconut genetic resources of Thailand In: CIRAD/IBPGR, ed. Report of a working session for implementation of the International Coconut Database. Montpellier: CIRAD/IBPGR
45. Uhl NW and Dransfield J 1987. Genera Palmarum. Lawrence: Allen Press
46. Whitehead RA 1966a Sample survey and collection of coconut germplasm in the Pacific Island. Overseas Research Publ. No. 16. London: Minisrty for Overseas Development
47. Whitehead RA 1966b Some notes on dwarf coconuts palms in Jamaica. Trop Agric 44:277-294
48. Whitehead RA 1968 Collection of coconut germplasm from the Indian/Malasysian Region, Peru and the Seychelles and testing for resistance to lethal yellowing disease in Jamaica. Rome: FAO
49. Zizumbo D 1993 El cultivo del cocotero en el Occidente de México: (Siglos XVI-XVIII). In: Juan de la Fuente *et al.*, eds. Agricultura y agronomia en México: 500 años, pp257-280. Chapingo: Universidad Autonoma de Chapingo
50. Zizumbo D, Hernádez F and Harries HC 1993 Coconut varieties in Mexico. Econom Bot 47:65-78
51. Zizumbo D and Arellano J 1993 Establecimiento de germoplasma de cocotero para su evaluación ente el amarillamiento letal. Reporte Tecnico No.1. CONACyT 0598-N9109. México

Safe movement of coconut germplasm

MICHEL DOLLET
CIRAD-CP, Laboratorie de Phytovirologie des Regions Chaudes, CIRAD-ORSTOM (LPRC), BP 5035, 34032 Montpellier Cedex 1, France

1. Introduction

Integrated plant disease control often involves a search for tolerant or resistant planting material. For tree crops that are grown for more than 20 years, such as coconut (palms are still being harvested after 50 or 60 years in many areas of the world), genetic control is a major component of integrated control. Sometimes it is the only solution, either because it is, for example, impossible to control the vector, or because other solutions are too costly. For instance, the only way to control oil palm vascular wilt is to plant resistant lines. In the case of lethal yellowing in Florida and in Jamaica, vector control has not been developed, and apart from their high cost, the limitations of the preventive antibiotic treatments used in Florida have become apparent over the years. Therefore, the major recommendation is to replant with tolerant ecotypes or hybrids, such as: 'Malayan Dwarfs', the 'Maypan' ('Malayan Dwarf' x 'Panama Tall') hybrid, 'Ceylon Dwarf', 'Red Spicata' and 'Fiji Dwarf' [12]. Countries that do not have access to large germplasm collections have to import as diverse a collection as possible of different ecotypes, and plant "varietal performance trials" to test disease tolerance. However, introduction of new material has to be made under controlled conditions to avoid the appearance of new diseases. Viruses have often appeared in new countries that have introduced seeds from abroad. This paper examines the risks involved with the introduction of new coconut germplasm.

2. Benefits of germplasm introduction

Plant germplasm has always been exchanged between regions: firstly to introduce new crops on different continents (*e.g.* potato in Europe, coffee in South America, oil palm in Southeast Asia, *etc.*), then to improve yields and quality and control

C. Oropeza et al. (eds.), Lethal Yellowing: Research and Practical Aspects, 139–147.

diseases. With regards to coconut, the work carried out in the Ivory Coast is a reliable reference. Since 1950, 53 different ecotypes have been introduced from all the tropical region (Latin America, Caribbean, Africa, Indian Ocean, Pacific and Far East). The genetic diversity of this collection has been studied and its vegetative and production characteristics measured many times. Based on the data obtained, systematic crossing of ecotypes (hybrids) has been undertaken [4]. To date, 84 single between-ecotype hybrids have been tested in the Ivory Coast, 64 of which have already been evaluated. Apart from five crosses, all of these hybrids have shown significantly higher production levels than the local ecotype, the 'West African Tall' (WAT). Yields were double for four of the hybrids, and yields 66% higher than those of WAT were obtained for 35 of them [4]. Some of these hybrids may be of interest for genetic control of diseases. The 'Panama Tall' x 'Malayan Yellow Dwarf' hybrid, which is tolerant of lethal yellowing, also appears to be tolerant of *Phytophthora*, which causes nut-fall. Yields of 5.4 t of copra/ha have been obtained at 10 years of age. Some 30 years after the first introduction of germplasm, no new diseases from other continents or other African countries have been observed in the Ivory Coast.

3. Theoretical risks of germplasm introduction

The theoretical risks of germplasm exchange are undeniable The most widely acknowledged risk is the introduction of viruses *via* seeds. Despite all the precautions taken on an administrative level by quarantine or plant protection services, it is impossible to totally prevent tourists, officials or diplomats entering a country with a few seeds intended for their gardens, or to prevent companies or private enterprises, on the lookout for new varieties, from importing seeds illegally. For example, in the early 1980s, peanut stripe virus found its way into the USA in seeds from China, despite all the regulations in force.

It is relatively easy to prevent the introduction of fungi, nematodes or insects using standard "seed-disinfection" methods, whereas it is difficult to guarantee that seeds are "virus free" or "viroid-free". Recently, Mink [8] reported a total of 108 plant viruses can be transmitted by seeds or pollen. In the same article, he also identified six viroids transmitted by seeds: apple scar skin, apple dapple, chrysanthemum stunt, coleus, hop stunt-cucumber strain and potato spindle tuber. These rough figures will have to be substantiated because several parameters are involved in virus or viroid transmission by seeds. Without claiming to be exhaustive, some of the parameters that are of prime importance will be discussed with respect to controlling germplasm exchange.

a) If a virus is detected in a seed, this does not necessarily indicate that the plant obtained from it will be contaminated. In fact, the only inevitable vertical transmission is through embryo contamination. However, it is virtually impossible to carry out diagnostic tests on embryos. Furthermore, not all virus-infected embryos produce virus-infected plants. There is also very little probability that viruses or viroids detected in the cotyledon, albumen or seed coat leads to transmission.

b) It is almost impossible to select, *a priori,* seeds that may or may not transmit a given virus, since transmission will depend on a combination of different parameters: the development stage reached by the plant at the time of infection, interactions between the virus and the plant (virus isolate, strain or group and plant variety, cultivar and species), and environmental factors, particularly temperature at the time of germination and during the first stages of plant growth.

c) The virus multiplication site: phloem viruses, for example, do not seem to be transmitted by seeds [8]. Certain virus groups are not transmitted, whereas others are, and there are only two groups whose members are all transmitted: cucumoviruses (none have been observed in oil palm or coconut) and cryptoviruses, but these viruses do not cause any symptoms in the host plant and have not yet been associated with any specific disease.

It is therefore important not to be alarmist. It is clear that through the constant improvement of detection techniques (ELISA, molecular hybridisation, amplification by polymerase chain reaction [PCR] and the latest development, immuno-PCR), it will be increasingly simple to detect the presence of viruses or viroids in one or another part of plants or seeds at a given stage of development. However, this does not mean that there will be transmission.

Little is known about the exact mechanisms involved with virus transmission by pollen. Nine viruses have been demonstrated to be transmitted by pollen (4 ilarviruses, 3 nepoviruses, 1 sobemovirus and 1 unidentified). To date, none of these groups have been observed in coconut palms. According to research on this topic, pollen could be involved in two types of virus transmission: horizontal, through contamination of the pollinated plant and vertical, through seed contamination. However, several authors have stated that embryos are physically separated from mother plant tissues by thick callose barriers. Furthermore, there is no continuity of the vascular bundles, particularly phloem, between the embryo and the plant [9]. This comment is also relevant for the transmission by seeds or pollen of mycoplasmalike organism (MLO) diseases. The same arguments are very widely accepted as proving that it is impossible for MLO diseases to be transmitted by seeds [11]. The other arguments put forward by McCoy [11] are based on the fact that MLO diseases generally cause fruit abortion or fall. This latter symptom is well known in the case of coconut lethal yellowing.

Lastly, to conclude this theoretical section, one cannot, and must not, forget certain self-evident truths. No sane breeder or agronomist will search for planting material in diseased zones, and most of the major pests and diseases affecting germplasm exchange show symptomatologies that are difficult to overlook (*e.g.* dwarfism, yellowing, proliferation, death). With the exception of cryptoviruses, and a few latent viruses (*Chenopodium amaranticolor*, hop latent viroid) which do not generally have any significant economic impact, there are minimal risks involved in collecting germplasm in healthy production areas where no major diseases are evident.

4. Introduction of new risk-free coconut germplasm

Three possible situations can be found in the zone of origin of germplasm:

a) there is a clearly identified disease, with known symptoms and etiological agent, hence a means of indexation,

b) there is a clearly defined disease in terms of symptomatology, but the pathogen is not known and there is no means of indexation, and

c) there is no disease.

If there is no clearly identified disease in the seed collection zone and coconut palms have been grown there for several decades (sometimes more than 100 years), there are few risks involved with importing seeds from that zone. Obviously, the old coconut palms (more than 50 years) which have never received fertiliser or phytosanitary treatments, although regularly subjected to attacks from leaf-eaters or other pests, often have symptoms of yellowing or stem shrinkage and are less productive than coconut palms with "maintenance". However, these deficiency symptoms on coconuts that have become spindly should not be confused with pathological or lethal symptoms such as those of lethal yellowing or cadang-cadang. In both cases, there are disease foci, an "epidemic" and an advancing. In any event, once again, breeders' common sense will prompt them to take fruit from the healthiest, most vigorous palms.

In the event of identified diseases, great care will have to be taken. However, not all diseases pose the same potential risks, depending on whether they are curable (numerous fungal diseases) or incurable (viroids, viruses, mycoplasmas), on whether the seed can be totally disinfected to remove any trace of pathogens or predators or not (presence of viruses or viroids in the embryo). The risks posed by each type of pathogen will therefore be evaluated.

In the case of diseases from unknown origin about which nothing is known of the pathogen and its eventual seed transmission, we must avoid importing material from the infected zone.

4.1 Insects and mites

Insects (disease vectors or predators) and mites can be controlled by insecticide treatments (remove the perianth from the nut and soak for 3 min in a solution of monocrotophos and an acaricide). In fact, the major risks of introducing insects or mites result from introducing whole plants or germinated nuts.

4.2 Fungi

Fungal diseases will not be discussed since: there are generally control methods available, they often occur world-wide (*e.g. Phytophthora*) and the risk of transmission by seeds is almost nil after classical fungicide treatments used for germplasm exchange (*e.g.* at the same time as the insecticide treatment, after removing the perianth, soak in a Folpel + thiophanate-methyl type solution).

The only fungal disease that requires particular attention is leaf blight, which has only been observed in Brazil and is known locally as "lixa pequena", "lixa grande" and "queima das folhas" [19]. This disease results from a combination of two types of fungus: *Catacauma torendiella* or *Coccostroma palmicola*, primary parasites on which *Botrysphaeria* sp. establishes itself, leading to the main type of damage, drying out of the leaves, and the subsequent fall in production.

4.3 *Viruses*

Only one virus has been identified in coconut to date, coconut foliar decay virus (CFDV) [14, 16]. This virus and disease have only been observed in Vanuatu. The disease is easy to detect due to a characteristic symptomatology. Although the virus has been identified in the nut husk, it has not yet been proved that it is transmitted by seeds. By virtue of its limited geographical distribution, its characteristic symptomatology and the fact that it is probably not transmitted by seeds, it is easy to avoid introducing this virus.

4.4 *Viroids*

Two very similar viroids have been identified in coconut: coconut cadang-cadang viroid (CCCVd) in the Philippines [13] and coconut tinangaja viroid (CTiVd) in Guam [3]. Both viroids are associated with pathological syndromes with specific symptoms (equatorial scarification of nuts, translucent bright yellow spots on the leaflets, diminishing inflorescences, *etc.*). Only one case of transmission by seeds (1/300) has been reported (quoted in [15]). CCCVd has been detected in nuts, pollen and embryos (frequencies not given) [6]. However, as stated above, this does not indicate that all fruit, or all pollen, will lead to diseased plants. Indeed, if this were the case, there would be more than one report of cadang-cadang transmission by seeds. As with CFDV, avoiding collecting nuts in the limited zones affected by these two diseases prevents any risk of introducing cadang-cadang.

4.5 *CCCVd viroid-like sequences*

Hanold and Randles [7] reported the existence of viroid-like sequences in samples of coconut and various palms (*Elaeis guineensis*, *Areca catachu*, *Metroxylon warburgii*) collected in various countries between Indonesia and Vanuatu. However, none of the coconut palms from which samples were taken showed typical symptoms of coconut cadang-cadang disease. It is important to note that these viroid-like sequences, identified by molecular hybridisation, have very variable electrophoretic migration patterns. The figures sometimes even show hybridisation throughout the electrophoresis gel, whereas CCCVd controls only show hybridisation in two distinct zones, corresponding to the two types of CCCVd molecules.

It is worth pointing out that our laboratory, as well as two other teams, have detected viroid-like molecules in oil palm in South America [1, 2, 10, 17]. These viroid-like molecules have also been found in healthy oil palm populations in West Africa (quoted in [7]).

The only risk of introducing viroids, pathogens specifically associated with a pathology, would therefore seem to be limited to the cadang-cadang extension zone in the Philippines, Guam and perhaps the areas with orange spotting in the Solomon Islands [5]. This last case requires confirmation by pathogenicity tests of the CCCVd viroid-like molecules found.

4.6 Mycoplasmalike organisms

For reasons mentioned previously (4.3), mycoplasmalike organisms (MLO), being intraphloemic organisms, are probably not transmitted by seeds. Furthermore, the lethal yellowing diseases that MLOs cause, are well known (USA, West Indies, Ghana, Togo, Nigeria, Cameroon, Tanzania, Mozambique) (Eden-Green, this volume; Jones *et al.*, this volume). A slow coconut decay found in India, Kerala root wilt, may also be a MLO disease [18]. Provided that the necessary precautions are taken to avoid introducing the vector of the MLO disease, the risks are nil.

*4.7 Trypanosomes (*Phytomonas*)*

Like MLOs, these pathogens are intraphloemic microorganisms that cause the decays, hartrot and Cedros wilt. As a result, they are not considered a major threat when transporting coconut germplasm.

4.8 Nematodes

Rhadinaphenlinchus cocophilus, which is responsible for red ring disease, is found throughout South and Central America. It is transmitted by the palm weevil,

Rhynchophorus palmarum. The conventional treatments to prevent the introduction of the vector, *Rhynchophorus*, are easily applied.

5. Conclusions

For tree crops, genetic resistance to diseases is an inevitable component of integrated disease control programs. Establishing performance trials requires the introduction of new ecotypes in order to: find ecotypes or hybrids (also using local ecotypes as a parent) that are disease tolerant or resistant, and plant high-yielding palms that justify the cost of establishing new plantations. To date, the countries that have implemented large-scale germplasm introduction programs have remained free from new exotic diseases. As a result of germplasm introduction, they have significantly increased coconut yields. Only virus (CFDV) and viroid diseases (CCCVd and CTiVd) pose a risk in germplasm importation. However, the distribution zones of these two diseases, cadang-cadang and coconut foliar decay, are well known, as is their symptomatology. Using common sense by not collecting seed-nuts in these zones, or from palms with suspicious symptoms, is an almost fail-safe way of preventing the introduction of new diseases.

Finally, to further reduce the residual risks associated with coconut exchange, and in the event of any doubt (no data on the effectiveness of phytosanitary products against various micro-organisms that may be associated with the nut), germplasm exchange in the form of *in vitro* embryos could be considered in future.

References

1. Beuther E, Wiese U, Lukacs N, Van Slobbe WG and Riesner D (1992) Fatal yellowing of oil palms: search for viroids and double stranded RNA. J Phytopath 136:297-311
2. Bernard V, Beuther E, Mazzolini L, Dambier D and Dollet M (1991) Caractérisation d'un ARN de type viroïde et détection de dsRNA dans le palmier à huile. In: 3ème rencontre de virologie végétale. Aussois: INRA/CNRS
3. Boccardo G, Beaver RG, Randles JW and Imperial JS (1981) Tinangaja and bristle top, coconut diseases of uncertain etiology in Guam and their relationship to cadang-cadang disease of coconut in the Philippines. Phytopath 71:1104-1107
4. Bourdeix R (1993) Résultats et perspectives de l'amélioration génétique du cocotier. In: Séminaire International sur le Cocotier- Recherches Européennes au service du cocotier, Sep 1993. Montpellier
5. Hanold D and Randles JW (1989) Cadang-cadang like-viroid in oil palm in the Solomon Islands. Plant Dis 73:183
6. Hanold D and Randles JW (1991) Coconut cadang-cadang disease and its viroid agent. Plant Dis 75:330-335

7. Hanold D and Randles JW (1991) Detection of cadang-cadang viroid-like sequences in oil and coconut palm and other monocotyledons in the South West Pacific. Ann Appl Biol 118:139-151
8. Mink GI (1993) Pollen and seed transmitted viruses and viroids. Annu Rev Phytopath 31:375-402
9. Mandahar CL (1981) Virus transmission through pollen and seed. In: K Maramorosch and KL Harris, eds. Plant Diseases and Vections. Ecology and Epidemiology, pp 241-249. New York: Academic Press
10. Mazzolini L, Dambier D and Dollet M (1989) Mise en évidence d'une molécule type viroïde dans le palmier à huile en equateur. In: 2ème Rencontre de Virologie Végétale. Aussois: CNRS-INRA
11. McCoy RE (1979) Mycoplasma and yellow diseases In: RF Whitcomb and JG Tully, eds. The Mycoplasmas Vol III. Plant and Insect Mycoplasmas, pp 229-264. New York: Academic Press
12. McCoy RE, Howard FW, Tsai JH, Donselman HM, Thomas DL, Basham HG, Atilano RA, Eskafi FM. Britt L and Collins ME (1983) Lethal Yellowing of Palms. Uni Florida Agric Expt Sta Bull No 834
13. Randles JW (1975) Association of two ribonucleic acid especies with cadang-cadang disease of coconut palm. Phytopath 65:163-167
14. Randles JW and Hanold D (1989) Coconut foliar decay virus particles are 20-nm icosahedra. Intervirol 30:177-180
15. Randles JW and Imperial JS (1984) Coconut cadang-cadang viroid. Description of plant viruses No 287. Kew: CMI/AAB
16. Randles JW, Julia JF, Calvez C and Dollet M (1986) Association of single-stranded DNA with the foliar decay disease of coconut palm (FDMT) in Vanuatu. Phytopath 76:889-894
17. Singh RP, Avila AC, Dusi AN, Boucher A, Trinidade DR, Van Slobbe WG, Ribeiro SG and Fonseca MEN (1988) Association of viroid-like nucleic acids with the fatal yellowing disease of oil palm. Fitopatol Bras 13:392-394
18. Solomon JJ and Govindakutty MP (1983) Association of mycoplasma-like organism with the coconut root (wilt) disease in India. J Plant Dis Protect 90:295-297
19. Souza Filho BF, Santo Filho HP and Robbs CP (1979) Etiologia da queima das fohlas do coqueiro. Fitopatol Bras 10:4-6

The genetics of durable resistance to lethal yellowing disease

HUGH C. HARRIES
International Coconut Cultivar Registration Authority, 17 Alexandra Road, Lodmoor Hill, Weymouth, Dorset DT4 7QQ, UK

1. Introduction

In 1984, the Food and Agriculture Organisation of the United Nations (FAO) held a meeting "to review the 'State of the Art' of disease resistance breeding in tropical perennial crops". Subsequently published as "Breeding for durable resistance in perennial crops", the proceedings dealt with banana, cacao, coffee, oil palm, rubber and sugar cane, but omitted the coconut palm, *Cocos nucifera* L. [6]. This is surprising: firstly, because FAO had a general involvement in coconut genetic resources [5], and secondly because FAO specifically encouraged a coconut breeding program for lethal yellowing disease resistance in Jamaica [25]. That program achieved practical results [11] and, despite localised losses, has had long-lasting benefits [14]. In setting the record straight, it is theorised that resistance to lethal yellowing disease in coconut is durable and a possible mode of inheritance is proposed. The apparently lower resistance to lethal disease in Tanzania can also be accommodated by the theory.

A theory of durable resistance is important for the Caribbean and Latin America where there is a chance to contain lethal yellowing disease by planting resistant material [17]. It is important in East Africa where lethal disease jeopardises the success of coconut replanting programs [18, 19, 20]. It is important in West Africa where similar diseases are known [3], or suspected [4], and also in the Indian sub-continent [21]. Even in Southeast Asia, where resistance probably evolved, MLOs occur in diseased palms [22].

Lethal yellowing disease of coconut and of about thirty other palm species in parts of the Caribbean, and lethal disease of coconut in East Africa, all implicate mycoplasmalike organisms (MLO). There are recent reviews of the similarities and differences between these and other diseases with superficially similar

C. Oropeza et al. (eds.), Lethal Yellowing: Research and Practical Aspects, 149–171.

symptoms [18, 19, 20]. In presenting data from Tanzania, those authors suggested that the different response of the same varieties tested in Jamaica and Tanzania was because the causal MLOs in the two regions were different. Comparable results for coconut MLO diseases elsewhere are not yet available.

When breeding coconut palms for resistance to lethal yellowing disease in Jamaica, Harries [7] pointed out that the requirements of an efficient selection method could not be met. The same constraints apply to lethal disease in Tanzania. It is not possible to subject enough test plants to a high infection pressure at a developmental stage in which symptom expression can be rapidly and confidently determined. In some situations, other pathogens produce similar symptoms (like Marchitez in South America). Even effects of *Phytophthora* bud rot, drought and waterlogging may be confused when no other reports precede or follow one-time sightings. Unavoidably, coconut planting density is low (80-140/ha), palms cannot be artificially infected and infection pressure depends on disease incidence in the immediate neighbourhood of the trial, over which there is no control. Symptoms take time to appear (3-6 months after infection in young palms and 7-15 months in mature palms) and a full syndrome of symptoms shows only in bearing palms. Indication of the presence of MLO is possible by fluorescence light microscopy or by tetracycline-mediated symptom remission, but confirmation requires electron microscopy. Chemotherapy and electron microscopy are too time-consuming and expensive to be used as routine screening procedures. DNA probes and PCR are a promising development [12, 13] (Harrison, this volume).

Both in Jamaica and in Tanzania, the screening method used is to replicate variety trials in, or close to, areas of diseased coconuts. Infection has to be left to chance. Monitoring the expression of symptoms, culminating in palm death, takes place at regular monthly intervals. The following results are from many years of field exposure trials in Jamaica and Tanzania.

The names given to coconut varieties and hybrids distinguish origins rather than relationships. Coconut populations are highly variable. Where they are sufficiently distinctive, local and regional populations receive names such as 'Jamaica Tall', 'East African Tall', 'Malayan Dwarf' or 'Pemba Dwarf'. Colour variants and local ecotypes within such populations also receive names. Names applied to individual coconut palms by domestic cultivators do not necessarily represent varieties but are merely the more interesting individuals in a population. There is an International Coconut Cultivar Registration Authority to clarify variety status [10].

2. Differences between breeding programs in Jamaica and Tanzania

Planning the coconut breeding program began in Jamaica in 1959 [24]. The first crosses, made in 1961, used locally available varieties. A particular parent, the 'Malayan Dwarf', had grown in diseased areas for more than ten years, so its high degree of resistance was recognised. However, there was no certainty that this resistance would be transmitted to hybrid progeny. Field exposure trials of local and hybrid material were a major part of the breeding program. After studying local material, exotic germplasm was introduced from a wide range of origins [23, 25]. Thereafter, all trials had local and introduced material side by side. Within fifteen years, variety differences in resistance were clear and an acceptable hybrid went into commercial production [11]. In the eighteen intervening years, the 'Maypan hybrid' and its 'Malayan Dwarf' seed parent have continued to give commercial yields in Jamaica and good resistance, both there and in Florida. Howard *et al.* [14] acknowledged this. Despite writing a pessimistic title ("Unusually high lethal yellowing disease incidence in 'Malayan Dwarf' coconut palms on localised sites in Jamaica and Florida"), these authors said:

> ". . . the relatively high incidence of LY in 'Malayan Dwarf' and 'Maypan' coconut palms was confined to limited localities. In frequent travels over the past decade in Florida and Jamaica we have observed many thousands of 'Malayan Dwarf' and 'Maypan' coconut palms and have seldom seen these affected by LY."(1)

In Tanzania, the coconut improvement program differed from that in Jamaica. The 'East African Tall' (EAT), 'Pemba Dwarf' (PD), locally occurring 'Malayan Dwarf', 'King Coconut' and EAT x PD hybrids were discounted. Instead, advice was given to use F_1 hybrids bred by the French-based Institut de Recherches pour les Huiles et Oléagineux (IRHO) at its Marc Delorme Station in Côte d'Ivoire. The IRHO breeding program concentrates on yield of copra [16] and assumes that resistance to MLO diseases is inherited from the 'Malayan Dwarf' seed parent. The IRHO hybrid seed unit is a commercial operation that sells hundreds of thousands of parental seed and kilogram quantities of pollen to multiplication units in other countries. In many of these countries, including Tanzania, coconut "breeding" was limited to making selections in the local tall population. These selections provided planting material against which the imported hybrids were expected to show

1 Dr. Howard, in a recent personal communication, stated that another paper is needed ". . . emphasizing the "normal" situation with the Malayan dwarf varieties - their resistance in general" and the ". . . massive numbers in Jamaica and Florida that held up under the LY epidemics in both countries".

advantage. In Tanzania, test quantities of 'East African Tall', and imported hybrids were planted in demonstration plots for that purpose. Despite a good start, the poor performance of the imported hybrids in the face of lethal disease and drought came as a shock. The hybrid seed production program in Tanzania harvested 200 000 seednuts in the first year of production (1985). Production rose to more than one million in 1987-88 but dropped to 100 000 in 1991. At that point the National Coconut Development Program (NCDP) stopped commercial-scale seed production. Hybrid seed continues to be produced as breeders' test material for on-farm trials. On the positive side, the failure of the introduced hybrids increased interest in local germplasm [15, 20] and the possibility of disease resistance breeding. Although NCDP Disease Control section field exposure trials included 'East African Tall' progenies as early as 1984, not until 1990 was a national breeding program started. This program aims to produce disease-resistant and drought-tolerant hybrids or true-breeding varieties, adapted to local conditions, productive despite low inputs but responsive to improved management.

There are other ways that the situation differs between Tanzania and Jamaica. The NCDP Disease Control Section imported varieties, similar to those that showed promise in Jamaica, and tested them in Tanzania in the expectation that they would have a satisfactory level of resistance. They did not. Nor does the 'Malayan Dwarf' have the same high resistance to lethal disease in Tanzania as it does to lethal yellowing disease in Jamaica. Yet neither has the lethal disease susceptibility of the 'East African Tall' variety in Tanzania been as high as the lethal yellowing disease susceptibility of the 'Jamaica Tall' in Jamaica. This may be due to a combination of environmental and genetic factors. The review by Schuiling *et al.* [20] covered the former. The latter is the subject of the remainder of this paper.

3. Experimental data

3.1 Field exposure trials

3.1.1 Lethal yellowing disease in Jamaica

The results shown in Tables 1 and 2, published in the Jamaica Coconut Industry Board annual research report [2] and made the basis of the paper by Harries [7], have not changed significantly. Earlier results are in earlier annual reports and in Whitehead [24]. Later results appeared in subsequent annual reports and in Been

[1]. The trial locations, well distributed through the coconut growing areas of Jamaica, involve different soil types and rainfall zones. The constraints, as pointed out elsewhere, are the limited number of plants at some sites and the necessity of relying on natural infection. However, lethal yellowing disease was very active at most sites from the late 1960s to the early 1980s. A subsequent decline in disease activity resulted from a combination of two factors that reduced the susceptible coconut population. The first was the high losses of the very susceptible 'Jamaica Tall', which left areas free of mature coconuts. The second was the replanting of these areas, under the very effective lethal yellowing disease rehabilitation scheme, with 'Malayan Dwarf' and the resistant 'Maypan' hybrid.

3.1.2 Lethal disease in Tanzania

The results in Table 3, published in Schuiling *et al.* [20], include epidemiological information. Similar constraints apply as to those mentioned previously for the Jamaica field exposure trials. There can be considerable disease activity at some sites but very little at others. There are localities in both countries where coconuts grow well and others where they grow badly because of the marginal amount of rainfall, its poor distribution, lack of ground water or absence of irrigation. Generally, the coconut-growing areas of Jamaica have a better total rainfall and better distribution than those in Tanzania. Such differences not only influence the coconut palms, but also the MLO, the vector and any alternate host-plants.

3.2 Categorisation of results

From these results, it is possible to categorise varieties by their reaction to lethal yellowing, as in the Jamaica data summarised by Harries [7] according to resistance or susceptibility. The present paper takes that categorisation a stage further (Table 4) in the light of new information on lethal disease in Tanzania. The categories in each part of Table 4 are based on the mean data for each variety across the various sites shown in Tables 1, 2 and 3. These values were initially listed in descending order of resistance. Some positions (indicated in the table by underlining & italics) were then adjusted, subjectively, to allow for differences attributed to plant age, disease activity at the trial site or sample size. Finally, arbitrary distinctions were made between five categories: highly-resistant, resistant, medium-resistant, susceptible, highly-susceptible.

Table 1. Incidence of lethal yellowing disease in local and introduced varieties [2].

Variety	Details	Number planted (No.) and mortality (%)											
		Fair Prospect		Woodstock		Kildare		Caenwood		Orange River		All Sites	
		No.	%	No.	%	No.	%	No.	%	No.	%	No.	%
Bougainville Tall	(a)	25	28	21	38	-	-	-	-	-	-	46	33
Cambodia Tall	(b)	27	30	-	-	-	-	-	-	48	6	75	15
Ceylon[1]Dwarf	(b)	34	3	-	-	-	-	-	-	-	-	34	3
Ceylon[1]Tall	(b)	18	94	-	-	20	25	48	73	55	16	141	43
Fiji Dwarf (Niu Leka)	(a)	54	44	39	36	10	80	-	-	-	-	103	35
Fiji Tall (& Niu Vai)	(a)	26	54	13	85	24	54	-	-	-	-	63	60
Indian Dwarf	(b)	9	0	-	-	-	-	-	-	-	-	9	0
Indian Tall	(b)	10	90	-	-	-	-	24	88	-	-	34	88
Jamaica Tall	(a)	49	92	74	88	280	88	48	81	48	31	499	82
Kar Kar Tall	(a)	12	66	20	35	-	-	-	-	-	-	32	47
King (ex Ceylon)	(b)	8	0	-		-	-	-	-	-	-	8	0
Malayan Dwarf (local)	(a)	304	1	42	0	139	8	96	10	29	0	610	4
Malayan Dwarf	(b)	42	2	-	-	-	-	19	0	-	-	61	1
Malayan Tall	(b)	54	41	-	-	-	-	96	38	146	4	296	22
Malayan Tall (ex BSIP)	(a)	49	51	20	45	20	10	-	-	-	-	32	50
Markham Valley Tall	(a)	12	58	20	45	-	-	-	-	-	-	32	50
New Hebrides[2]Tall	(a)	49	98	11	82	-	-	-	-	-	-	60	95
Panama Tall (local)	(a)	-	-	-	-	141	45	-	-	-	-	141	45
Peru Tall	(b)	-	-	-	-	-	-	48	27	-	-	48	27
Rangiroa Dwarf	(a)	24	92	6	67	-	-	-	-	-	-	30	87

Rangiroa Tall	(a)	24	67	-	-	7	29	-	-	-	-	31	58
Rennell Tall	(a)	13	62	50	62	-	-	-	-	-	-	63	62
Rotuma Tall	(a)	5	60	-	-	15	13	-	-	-	-	20	25
Samoa Tall	(a)	49	71	21	50	-	-	-	-	-	-	69	65
Sarawak Tall	(b)	19	47	-	-	-	-	48	50	19	5	86	40
Seychelles Tall	(c)	20	75	-	-	17	5	19	15	-	-	56	46
Solomon Tall	(a)	25	64	20	35	-	-	-	-	-	-	45	51
Tahiti Tall	(a)	49	82	20	70	28	50	-	-	-	-	97	70
Thailand Tall	(b)	50	30	-	-	-	-	48	40	58	3	156	24
Tonga Tall	(a)	6	83	-	-	20	25	-	-	-	-	26	38
Yap Tall (Thifow)	(c)	20	65	-	-	23	4	20	10	-	-	63	38

(a) CIB/ODA introduction. Planted 1965; except Jamaica Tall and Malayan Dwarf at Caenwood and Orange River (1967)
(b) CIB/FAO introduction. Planted 1967; except Ceylon Tall at Kildare
(c) CIB/FAO introduction. Planted 1969
(d) CIB/FAO introduction. Planted 1970

[1]Ceylon = Sri Lanka
[2]New Hebrides = Vanuatu

Table 2. Incidence of lethal yellowing disease in some local and F_1 hybrid coconut varieties. Based on [2].

Variety	Number planted (No.) and mortality (%)													
	Fair Prospect		Woodstock		Kildare		Caenwood		Rodney Hall		Gray's Inn		All Sites	
	No.	%	No.	%	No.	%	No.	%	No.	%	No.	%	No.	%
Jamaica Tall (JT)	49	92	126	83	311	86	48	81	60	98	-	-	594	87
Malayan Dwarf (MD)	304	1	140	9	139	9	96	10	306	5	180	11	1165	6
Panama Tall (PT)	-	-	-	-	141	45	-	-	-	-	-	-	141	45
Fiji-Malayan x JT	24	4	-	-	61	5	-	-	-	-	-	-	85	5
Fiji-Malayan x self	24	8	-	-	-	-	-	-	-	-	-	-	24	8
JT x Cambodia Tall	15	53	-	-	-	-	-	-	-	-	-	-	15	53
JT x Cameroon Dwarf	24	8	-	-	20	10	-	-	-	-	-	-	44	9
JT x Cuban Dwarf	-	-	-	-	37	65	-	-	-	-	-	-	37	59
JT x Malayan Dwarf	104	49	50	32	53	9	-	-	-	-	-	-	207	35
JT x Mozambique Tall	19	53	-	-	-	-	-	-	-	-	-	-	19	53
JT x Panama Tall	84	54	-	-	35	54	-	-	-	-	-	-	119	48
JT x Rennell Tall	24	92	-	-	-	-	-	-	-	-	-	-	24	92
JT x Solomon Tall (a)	-	-	-	-	-	-	-	-	-	-	-	-	36	25
JT x Tahiti Tall	14	79	-	-	-	-	-	-	-	-	-	-	14	79
MD x Cambodia Tall	7	29	-	-	-	-	2	50	-	-	-	-	9	33
MD x Cameroon Dwarf	48	15	-	-	-	-	-	-	-	-	-	-	48	15
MD x Jamaica Tall	87	23	50	26	60	20	130	29	60	7	526	18	913	20
MD x Mozambique Tall	6	17	-	-	-	-	4	50	-	-	-	-	10	30
MD x Panama Tall)	23	13	50	12	67	7	39	18	60	12	-	-	239	12
MD x Rennell Tall	24	29	-	-	44	25	16	44	60	28	-	-	144	29

MD x Solomon Tall	21	14	-	-	-	-	33	9	-	-	-	-	54	11
MD x Tahiti Tall	5	0	-	-	-	-	3	100	-	-	-	-	8	38
PT x Cameroon Dwarf	25	28	-	-	-	-	-	-	-	-	-	-	25	28
PT x Jamaica Tall	36	58	-	-	35	43	-	-	-	-	-	-	71	51
PT x Malayan Dwarf	-	-	50	12	47	15	-	-	-	-	-	-	97	13
PT x Mozambique Tall	3	33	-	-	-	-	-	-	-	-	-	-	3	33
PT x Rennell Tall	25	54	-	-	-	-	-	-	-	-	-	-	24	54
PT x Solomon Tall	23	17	-	-	-	-	-	-	-	-	-	-	23	17
PT x Tahiti Tall	3	0	-	-	-	-	-	-	-	-	-	-	3	0

(a) Orange River

Table 3. Incidence of lethal disease in some local and imported coconut varieties and F_1 hybrids in Tanzania. After Schuiling *et al.* [20].

Variety	Year planted	All		Kifumangao		Chambezi		Sotele		Pongwe	
		No.	% loss	No.	% loss	No.	% loss	No.	% loss	No.	% loss
Malayan Yellow Dwarf (MYD)	81	439	27	121	71	118	15	133	5	67	10
Malayan Red Dwarf (MRD)	81	244	26	64	69	55	9	63	13	62	10
Cameroon Red Dwarf (CRD)	81	237	37	62	90	59	25	62	19	54	7
Equatorial Green Dwarf (EGD)	81	224	41	63	98	55	29	62	16	44	9
Sri Lanka Green Dwarf (SGD)	83	154	26	60	60	-	-	64	5	30	7
Pemba Red Dwarf	85	145	23	66	49	39	5	-	-	40	5
King Coconut	88	60	28	30	53	30	3	-	-	-	-
East African Tall (LBS)	80	99	14	-	-	-	-	-	-	99	14
East African Tall (Mafia)	81	191	62	64	92	63	51	64	44	-	-
East African Tall (Chambezi)	83	419	34	62	82	224	25	66	23	66	33
West African Tall (WAT)	81	253	59	64	100	60	42	63	25	66	68
WAT x Rennell Tall	81	329	57	64	100	192	49	63	38	10	60
Vanuatu Tall	82	255	44	64	83	60	30	65	20	66	41
Rennell Tall	83	154	53	53	91	40	43	65	25	-	-
Polynesian Tall	83	163	61	60	95	40	45	63	40	-	-
Malayan Tall	83	185	53	61	92	58	43	66	26	-	-
Thailand Tall	84	230	46	59	66	57	35	63	59	51	16
Cambodia Tall	87	159	49	58	93	60	30	-	-	41	15

Panama Tall	89	141	31	49	43	50	36	42	10	-	-
Karkar Tall	89	121	27	50	34	50	32	21	0	-	-
Tagnanan Tall	89	119	24	48	25	50	24	21	24	-	-
Rotuma Tall	89	121	22	50	28	50	20	21	10	-	-
MYD x West African Tall	81	315	65	124	100	61	41	64	23	66	65
MYD x Polynesian Tall	82	263	60	63	98	70	60	64	22	66	60
MYD x Vanuatu Tall	82	258	36	64	84	63	33	65	12	66	17
MYD x Malayan Tall	82	254	46	64	95	58	41	66	17	66	30
MYD x East African Tall	83	161	53	55	93	44	14	62	45	-	-
MRD x West African Tall	81	180	48	64	98	61	21	63	22	-	-
MRD x Polynesian Tall	82	252	49	64	89	56	36	66	21	66	50
CRD x West African Tall	81	256	57	64	96	63	49	63	14	66	68
CRD x Malayan Tall	83	124	54	56	98			68	18	-	-
CRD x Rennell Tall	83	192	60	64	97	62	53	66	32	-	-
CRD x East African Tall	83	246	54	61	84	63	46	63	41*	59	44
EGD x West African Tall	81	259	52	65	99	60	42	63	19	71	47
EGD x Vanuatu Tall	82	140	37	27	82	48	42	65	15	-	-
EGD x East African Tall	83	237	54	61	90	50	30	63	57	63	37
WAT x Sri Lanka Green Dwarf	84	211	55	64	81	85	33	63	57	-	-

3.2.1 Jamaica

As previously reported, the most notable member of the highly-resistant group is the 'Malayan Dwarf', in all three colour forms. This is highly resistant and, being 80-90% self-pollinating and colour-marked, it largely breeds true-to-type. Four varieties are in Category 3, despite higher losses, because of a supposed common origin [8, 9]. These are the 'Panama Tall' (a local selection introduced to Jamaica in 1915), the 'Malayan Tall' ex BSIP (from the Solomon Islands), and the 'Kar Kar Tall' and 'Markham Valley Tall' (from Papua New Guinea). Also, the 'Rennell Tall' from the Solomon Islands is in Category 5, although its other features suggest it might be placed in Category 3. In the less resistant group of varieties, there have been no continued losses which would have reassigned some varieties into the highly-susceptible group in Category 5. Comparison with the highly-susceptible group suggests that real differences in levels of resistance do exist.

Among the F_1 hybrids, the categorisation depends on whether resistance comes from one or both parents. The 'Maypan' (Category 2) is a cross between 'Malayan Dwarf' (Category 1) and 'Panama Tall' (Category 3). It is better than 'Mayjam' (Category 3), which is a cross between 'Malayan Dwarf' (Category 1) and 'Jamaica Tall' (Category 5). This, in turn, is better than the unnamed Category 4 cross between 'Panama Tall' (Category 3) and 'Jamaica Tall' (Category 5). The reciprocal crosses behave similarly.

3.2.2 Tanzania

The Tanzanian results are categorised here for the first time. They are not so straightforward as those for Jamaica. In part, this is probably due to the environmental factors mentioned above. However, there could be a very real difference due to additional genetic factors.

To begin with, three varieties, included in Category 4, show high losses at a comparatively early age. These are the 'Panama Tall', the 'Kar Kar Tall' and the 'Cambodia Tall'. Like much other test material in Tanzania, these all came from a collection held in the Côte d'Ivoire and not from their respective countries of origin. Their recent plantings, in 1987 and 1989, are at trial sites where disease activity has increased following the repeated earlier planting of susceptible material. It is important to bear this factor in mind when comparing results between sites or within sites at different times. This was so in Jamaica, where repeatedly replanting susceptible 'Jamaica Tall' palms kept disease active at trials.

Most interestingly, the three 'East African Tall' accessions appear in separate categories which reflect real differences. The accession from the Livestock Breeding Station (LBS) in Category 2 is from a site in Tanga. Selection for resistance may have occurred there, where disease has been more active in the past [18]. The Chambezi accession in Category 3 is from a site where disease was active at the time of seednut collection, so the population would have contained some susceptible palms. The Mafia accession in Category 4 comes from a location where there was no disease prior to seednut collection and therefore no selection for resistance. Because of its disease freedom (prior to 1989), the Mafia location was also the source of seed and pollen sent to Côte d'Ivoire. The various imported hybrids featuring 'East African Tall' came from there.

At first sight, it might seem unsatisfactory that Category 3 and Category 4 of the lethal disease section of Table 4 should contain dwarf, tall and hybrid varieties. It might be thought that they should be in separate categories, as was possible for lethal yellowing disease. Indeed, the authors of the original paper each express a different preference for the position of some varieties in this section of Table 4. In fact, the exact positions are unlikely to be determined from the data at hand. It can be argued that there is no one exact position while there are unavoidable restrictions such as small populations, small plots and uncontrolled infection rates.

The small plot sizes and sporadic nature of the disease confuse the issue. If, as suggested later, two genes control resistance, then the overall resistance for any of this material would be 50%, putting all accessions into Category 3. However, results might range from 0 to 100% in samples of some populations; whereas others, with the same overall mean, would range only from 25 to 75% (see Fig. 2).

3. Discussion

3.1 Durable resistance hypothesis

There is a suggestion (Löhr pers. comm.) that the resistance mechanism is different for both diseases. That is immaterial to the present argument, since different genes controlling different resistance mechanisms can be present in the same variety.

Table 4. Coconut varieties and F1 hybrids categorised by the degree of resistance and susceptibility to lethal yellowing disease in Jamaica and lethal disease in Tanzania.

Category	Lethal yellowing results, Jamaica		Lethal disease results, Tanzania	
	Variety	Losses (%)	Variety	Losses (%)
Category 1	Indian Dwarf	0		
	King (ex Ceylon)	0		
Highly Resistant	Malayan Dwarf	1-6	None	
	Ceylon Dwarf	3		
	Fiji-Malayan x self	8		
Category 2	Fiji-Malayan x JT	5	East African Tall (EAT LBS)	14
	MD x Solomon Tall	11	Malayan Dwarf (MRD & MYD)	26-27
Resistant	MD x Panama Tall	12	Sri Lanka Green Dwarf (SGD)	26
	PT x Malayan Dwarf	13	Pemba Dwarf	23
	MD x Cambodia Tall	33	King Coconut	28
			Cameroon Dwarf (CRD)	37
Category 3	Cambodia Tall	15	MYD x VAT	17
	MD x Cameroon Dwarf	15	Equatorial Green Dwarf (EGD)	41
	PT x Solomon Tall	17	EAT (Chambezi)	34
	MD x Jamaica Tall	20	Vanuatu Tall (VAT)	44
	Malayan Tall	22	MYD x MLT	48
	Thailand Tall	24	MRD x WAT	48
	Rotuma Tall	25	EGD x WAT	52
	Peru Tall	27	MRD x PYT	49
Medium Resistant	PT x Cameroon Dwarf	28	CRD x WAT	57
	MD x Rennell Tall	29	WAT x RLT	57
	MD x Mozambique Tall	30		
	Bougainville Tall	33	West African Tall (WAT)	59
	Fiji Dwarf (Niu Leka)	35	Rotuma Tall	22
	JT x Malayan Dwarf	35	EAT (Mafia)	62
	MD x Tahiti Tall	38	Thailand Tall	46
	Tonga Tall	38	Rennell Tall (RLT)	53

	Yap Tall (Thifow)	38	MYD x EAT	53
	Sarawak Tall	40	Malayan Tall (MLT)	53
	Panama Tall (PT)	45	MYD x WAT	65
	Karkar Tall	47	MYD x PYT	60
	Malayan Tall (ex BSIP)	50	Tagnanan Tall	24
	Markham Valley Tall	50	CRD x EAT	54
	Rennell Tall	62	CRD x MLT	54
			EGD x EAT	54
Category 4	*PT x Tahiti Tall*	0	CRD x RLT	60
	JT x Cameroon Dwarf	9	Tahiti Tall (PYT)	61
More susceptible	*JT x Solomon Tall*	25	WAT x SGD	55
	PT x Mozambique Tall	33	*Panama Tall*	31
	JT x Panama Tall	48	*Karkar Tall*	27
	PT x Jamaica Tall	51	*Cambodia Tall*	49
Category 5	*Ceylon Tall*	43		
	Seychelles Tall	46		
	Solomon Tall	51		
	JT x Cambodia Tall	53		
	JT x Mozambique Tall	53		
	PT x Rennell Tall	54		
	Rangiroa Tall	58		
Highly susceptible	JT x Cuban Dwarf	59		
	Fiji Tall (& Niu Vai)	60	None	
	Samoa Tall	65		
	Tahiti Tall	70		
	JT x Tahiti Tall	79		
	Jamaica Tall (JT)	82-87		
	Rangiroa Dwarf	87		
	Indian Tall	88		
	JT x Rennell Tall	92		
	New Hebrides Tall	95		

As mentioned above, it must be emphasised that coconut varieties are essentially populations. In any population, most individuals resemble the norm but almost all vary in one or more respects. It would be difficult to expect to find a single resistant palm in a generally highly susceptible population. Yet, that is how the first resistance must have developed.

3.2 Theory of evolution of resistance

The coconut palm may have evolved as a wild plant on small uninhabited islands but domestication was on the continental coast and larger islands of Southeast Asia [8, 10]. Diseases that would be unlikely to spread from island to island in an archipelago could reach epidemic proportions on a continental coast. The wild type palms would be susceptible to diseases against which domestic types, selected by exposure and survival, would be resistant. The latter would show high (Category 1) resistance in self-pollinating varieties but intermediate (Category 3) resistance in out-crossing varieties. Such naturally developed resistance would be durable. Disease outbreaks would become rare events, stimulated by the introduction of susceptible coconut genotypes [9, 22] or the emergence of a more virulent strain [14]. However, when susceptible and resistant material grow together, variability of all characteristics, including disease resistance, will be generated by subsequent generations of introgressive hybridisation. This, together with small sample size, is the reason that the populations under test do not always give clear-cut results.

Tall and dwarf coconut varieties are very often contrasted and said to be respectively cross-pollinating and self-pollinating. This is an over-simplification. In practise, tall varieties sometimes self-pollinate between successive inflorescences on the same palm. Nor is there any mechanism to prevent dwarf varieties from cross-pollinating. This happens when they grow in small numbers, in proximity to other, usually tall, palms.

In the following estimates of resistance, the percentages mentioned suggest the proportion of the population that may be expected to die if exposed to infection. The estimate says nothing about the speed with which the palms might die or the time taken to reach the estimated level of loss, since the environment influences these factors.

3.3 *Lethal yellowing* L_1

From the results in Jamaica it is possible to consider resistance as if it were due to a single gene (Fig. 1). Its alleles are homozygous dominant (L_1L_1) in the resistant 'Malayan Dwarf' or homozygous recessive (l_1l_1) in the 'Jamaica Tall' and other susceptible varieties. Dominance is incomplete and the F_1 hybrid between 'Malayan Dwarf' and 'Jamaica Tall' is heterozygous (L_1l_1). It has intermediate resistance, approximating to fifty percent. Varieties such as the 'Panama Tall', that also show an intermediate degree of resistance, behave as if heterozygous, but only about half the palms have this genetic pattern. A quarter are homozygous dominant and the remaining quarter, homozygous recessive. Therefore, the F_1 hybrid between 'Malayan Dwarf' and 'Panama Tall' is not a simple heterozygote but, instead, carries practically equal proportions of the homozygous double dominant and the heterozygous forms. It therefore has a better degree of resistance, approximating to seventy-five percent.

Based on this explanation, it is possible to breed highly resistant tall varieties by crossing survivors (assumed resistant) and selecting from their progeny. However, eliminating the twenty-five percent susceptible segregants and the fifty-percent heterozygous segregants makes this program difficult to carry out. It must wait until disease screening can be done by inoculation or biochemical tests rather than by field exposure.

It also should be borne in mind that the simple segregation ratios presume there is no disease present. When susceptible palms die of disease the remaining resistant palms, both homozygous and heterozygous, will cross to give progeny with better resistance. This is how resistant varieties must have originally developed over many generations, depending on the number of genes actually involved. The fact that some 'Malayan Dwarfs' die and that tall and hybrid varieties show a range of intermediate resistances suggests that other genes may modify the expression of L_1. One of these might be L_2 (Fig. 2).

3.4 *Lethal disease* L_2

For lethal disease in Tanzania, it is suggested that a completely resistant palm is homozygous double dominant ($L_1L_1L_2L_2$). The same argument applies to resistance to any postulated new strain of the lethal yellowing pathogen. Since no variety tested in Tanzania is highly resistant, it seems that none carry this full

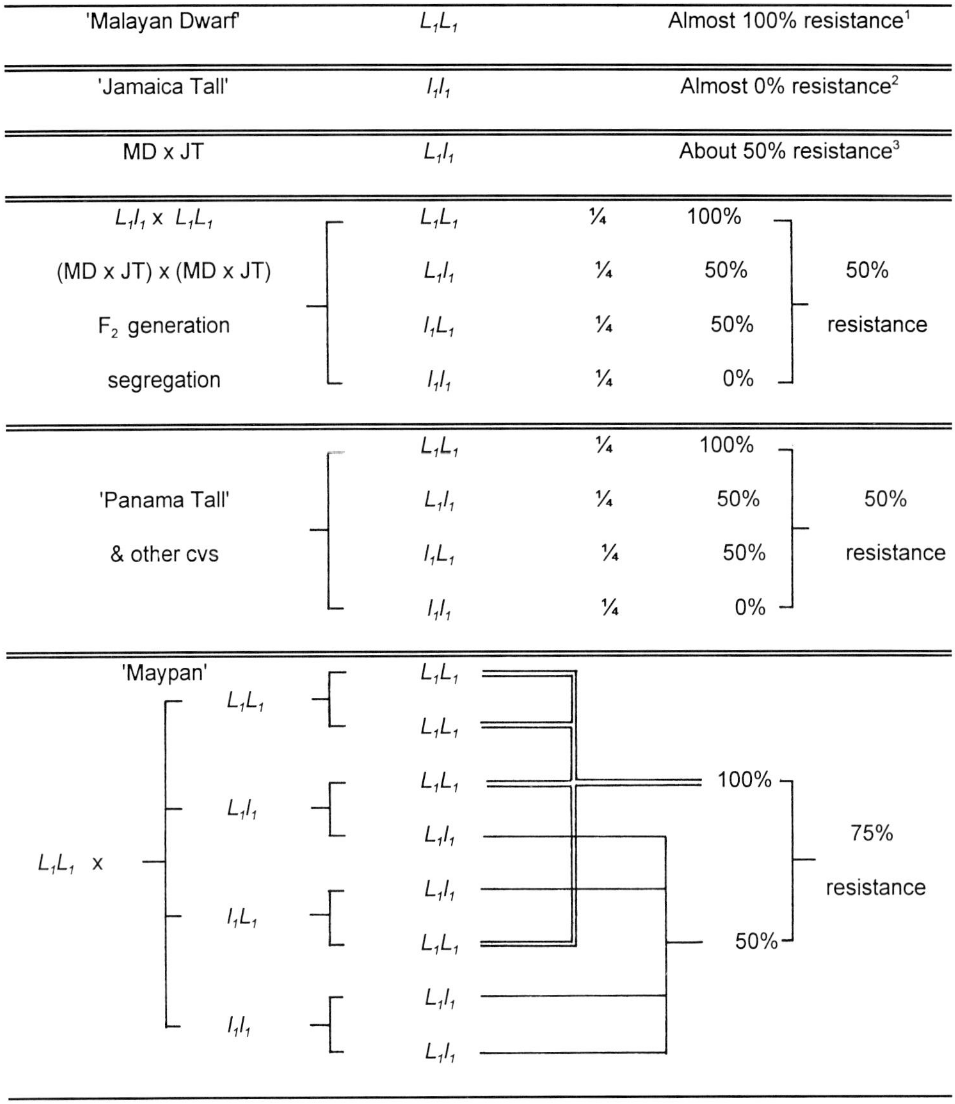

Fig 1. Single gene hypothesis of lethal yellowing disease resistance. [1]Resistance maintained by a high degree of self-pollination. [2]Susceptibility maintained despite a high degree of cross-pollination. [3]L_1 is incompletely dominant to l_1 and the heterozygote has intermediate resitance.

complement. Therefore, the dwarf varieties, largely self-pollinated, must be $L_1L_1l_2l_2$ and fifty percent resistant to lethal disease. Even the 'Rangiroa Dwarf' that is susceptible to lethal yellowing disease in Jamaica would be $l_1l_1L_2L_2$ and also fifty percent lethal disease-resistant if tested in Tanzania. Some varieties, either dwarf or tall, with a $l_1l_1l_2l_2$ genotype might be totally susceptible to both diseases. There are some accessions that are totally susceptible at some, but not all, sites.

The local tall varieties are populations in which individual palms are homozygous dominant, heterozygous or homozygous recessive for L_1 and L_2 in all possible combinations. Individual palms in a population may range from totally susceptible to totally resistant but, on average, populations (varieties) exhibit only fifty percent resistance. The explanation of better performance of the 'East African Tall' accession at the Livestock Breeding Station (LBS) is the elimination of susceptible palms by exposure to disease prior to seed selection. Hybrids between dwarf and intermediate forms are also fifty percent resistant but do not contain any totally resistant or totally susceptible individual palms (see Fig. 2).

'Malayan Dwarf'	$L_1L_1l_1l_1$	Approximately 50% resistance[1]		
Susceptible	$l_1l_1l_2l_2$	Almost 0% resisitance[2]		
Resisitant	$L_1L_1L_2L_2$	Almost 100% resistance		
		x Intermediate		
Intermediate	L_1L_2	L_1l_2	l_1L_2	l_1l_2
L_1L_2	100%	75%	75%	50%
L_1l_2	75%	50%	50%	25%
l_1L_2	75%	50%	50%	25%
l_1l_2	50%	25%	25%	0%
		x Intermediate		
Dwarf	L_1L_2	L_1l_2	l_1L_2	l_1l_2
L_1L_2	75%	50%	50%	25%

Fig. 2. Two gene hypothesis for lethal disease resitance. The same theory may also account for loss of resistance to lethal yellowing disease. [1]Resistance stays at 50% despite a high degree of self-pollination. [2]Susceptibility maintained despite a high degree of cross-pollination.

The overall effect is that the same varieties respond differently in lethal yellowing disease and lethal disease trials. It would be unusual to find high total resistance to lethal disease (and perhaps high total susceptibility, although small samples at some active disease sites may appear so). It might be possible to predict the genetic configuration for some varieties under test in Tanzania and therefore work out the best potential combinations for hybrid production. However, this would still depend on being able to identify the homozygotes in a segregating population. For the same reason, it would be possible, but even more difficult, to develop highly resistant true-breeding varieties. There may be other genes that modify the expression of L_1 and L_2. One of these might be L_3, but that possibility remains to be explored.

One further important point is the need to screen enough palms. Given the nature of the coconut seednut, and the cost of introducing germplasm and establishing field trials, small plots are unavoidable. The number of palms exposed at the seven locations in Jamaica averaged 22 and ranged from 3 to 526. In the four Tanzanian locations the average number of palms per variety was 52, ranging from 10 to 224. Almost unavoidably, small plots of segregating populations will give inconsistent results. Besides any ecological factors, this is why the varieties under test seem less susceptible in some places and more susceptible in others. To have any chance of recovering the homozygote within a segregating population for a single gene, the population can be quite small; ten plants give a ninety-five percent chance of recovery in 3:1 segregation. For two genes, assuming no linkage, the number rises to nearly fifty plants (15:1) while with three genes, it is almost 200 (63:1). If two genes control lethal disease resistance, then only about three palms from an original plot of fifty-two will carry the homozygous double dominant ($L_1L_1L_2L_2$). The simple situation outlined above is likely to be more complicated, and not just by further additive genes. Bear in mind that the resistant individuals also must survive hurricanes in the Caribbean, droughts in East Africa and other hazards common to coconut growing everywhere.

4. Conclusions

In suggesting that the two MLO must be different [20], other differences between East Africa and the Caribbean are not ignored. The growing conditions are different. The natural history of coconuts is different. They have been present for at least two thousand years in East Africa, but for less than five hundred in West Africa and America. The germplasm under test is similar but not identical. The vector is not known in Tanzania and the one that occurs in the Caribbean is not present in East Africa. Until any of these factors can account for the difference in resistance, it must be presumed that it is due to the genetic make-up of the coconut germplasm and the MLO.

In Jamaica, resistance has worked for the best part of fifty years and can surely be considered durable resistance, even for a perennial plant. Losses within the resistant material occurred from the start and even now remain limited to certain sites [14]. Nevertheless, this represents a potential loss of resistance and a threat to indefinite success. The threat from this disease became even more serious when it reached the coconut growing areas of mainland America (particularly México, but also Florida in the USA).

The fact that the Tanzanian results have not duplicated Jamaica results is also cause for concern. In Tanzania, there is a slight indication of resistance, and this is in certain local material. Even so, in the face of lethal disease for the last forty to fifty years, Tanzanian farmers continue to grow the 'East African Tall' coconut variety. This also is evidence of durable disease resistance in the coconut.

It should be remembered, however, that in both diseases, the surviving palms may simply have escaped infection. If they are resistant, selection within any tall population will be difficult. Depending on the frequency of the *L* alleles, resistance is likely to be heterozygous, due to cross-fertilisation. Any attempt to produce a uniformly highly resistant, cross-pollinating variety will be possible only if there is access to large backcross populations and the means by which to screen them. Harries [7] described how large backcross progenies could be produced by controlled pollination. There are reasons why this is not done. For instance, in Jamaica, there is nowhere to plant extensive new field exposure trials. Also, the agricultural success of the lethal yellowing disease rehabilitation program has reduced infection pressure because resistant material is now planted in once highly diseased areas.

The newly developed ability to screen coconut populations for MLO and, therefore, for factors contributing to disease resistance [12, 13] may resolve the problem of screening extensive segregating progenies. It is hoped that this hypothesis for genetically controlled resistance to lethal yellowing disease and lethal disease will help breeders and pathologists to make the preliminary identification of material suitable for testing.

Acknowledgements

This paper was based on a joint paper written with Alois Kullaya of National Coconut Development Programme (NCDP), Tanzania, Basil Been of Coconut Industry Board (CIB), Jamaica and Mark Schuiling (who like myself has worked with both CIB and NCDP). That paper, entitled "Selecting and breeding coconut palm (*Cocos nucifera* L.) for durable resistance to disease" has not yet been published.

References

1. Been BO (1981) Observations on field resistance to lethal yellowing in coconut varieties and hybrids in Jamaica. Oléagineux 36:9-11
2. CIB (1972) Botany/Plant Breeding. 14th Annu Res Rpt Coconut Industry Board Jamaica
3. Dabek A, Johnson CG and Harries HC (1976) Mycoplasmalike-organisms associated with kaincope and Cape St. Paul wilt disease of coconut palms in West Africa. PANS 22:357-8
4. Ekpo EN and Ojomo EE (1990) The spread of lethal coconut diseases in West Africa: incidence of Awka disease (or bronze leaf wilt) in the Ishan area of Bendel State of Nigeria. Principes 34:143-146
5. FAO (1969-81) Yearly Progress Reports on Coconut Breeding. In: FAO Technical Working Party on Coconut Production, Protection and Processing. Rome: Food and Agriculture Organisation
6. FAO (1986) Breeding for durable resistance in perennial crops. FAO Plant Production and Protection Paper 70. Rome: Food and Agriculture Organisation
7. Harries HC (1973) Selection and breeding of coconuts for resistance to diseases such as lethal yellowing. Oléagineux 28:395-398
8. Harries HC (1978) The evolution, dissemination and classification of *Cocos nucifera*. Bot Rev 44:265-320
9. Harries HC (1980) Malayan coconuts in the Caribbean: Caribbean coconuts in Malaysia. In: Proc Int Conf Cocoa Coconuts, 1978, pp 508-510. Kuala Lumpur
10. Harries HC (1991) The ISHS Coconut Registration Authority. Principes 35:154-155
11. Harries HC and Romney DH (1974) Maypan: an F_1 hybrid coconut variety for commercial production. World Crops 26:110-111
12. Harrison NA, Bourne CM, Cox RL, Tsai JH and Richardson PA (1992) DNA probes for detection of mycoplasmalike organisms associated with lethal yellowing disease of palms in Florida. Phytopath 82:216-224
13. Harrison NA, Richardson PA, Jones P, Tymon AM, Eden-Green SJ and Mpunami AA (1994) Comparative investigation of MLOs associated with Caribbean and African coconut lethal decline diseases by DNA hybridization and PCR assays. Plant Dis 78:507-511
14. Howard FW, Atilano R, Barrant CI, Harrison NA, Theobold WF and Williams DS (1987) Unusually high lethal yellowing disease incidence in Malayan Dwarf coconut palms in localized sites on Jamaica and Florida. J. Plantation Crops 15:86-100
15. Krain E, Issa JA, Kullaya A and Harries HC (1994) The coconut in Africa 2. The Pemba Dwarf in Zanzibar. Palm Enthusiast 10:9-15
16. Nucé de Lamothe M de, Wuidart W and Rognon F (1980) Premier bilan de 12 années de recherches genetiques sur le cocotier en Côte d'Ivoire. Oléagineux 35:131-144
17. Robert ML and Zizumbo D, eds (1990) La problematica del amarillamiento letal del cocotero en Mexico. Mérida: Centro de Investigación Científica de Yucatán
18. Schuiling M, Kaiza DA and Mpunami A (1992) Lethal disease of coconut palm in Tanzania. 2. History, distribution and epidemiology. Oléagineux 47:516-522
19. Schuiling M and Mpunami A (1992) Lethal disease of coconut palm in Tanzania. 1. Comparison with other diseases in East Africa. Oléagineux 47:511-515
20. Schuiling M, Mpunami A, Kaiza DA and Harries HC (1992) Lethal disease of coconut palm in Tanzania. 3. Poor resistance of introduced varieties. Oléagineux 47:693-698
21. Solomon JJ (1992) Recent advances in research on root (wilt) disease of coconuts. Indian Coconut J 22:11-17 [reprinted from J Plantation Crops 19:(2), 1991]
22. Turner PD, Jones P and Kenten RM (1978) Coconut stem necrosis: a disease of hybrid and Malayan Dwarf coconuts in north Sumatra and peninsular Malysia. Perak Planters' Assoc J pp 33-46
23. Whitehead RA (1966) Sample survey and collection of coconut germplasm in the Pacific islands (30 May-5 September 1964). Ministry of Overseas Development. London: HMSO
24. Whitehead RA (1968) Selecting and breeding coconut palms (*Cocos nucifera* L.) resistant to lethal yellowing disease. Euphytica 17:81-101

25. Whitehead RA (1968) Collection of coconut germplasm from the Indian/Malaysian Region, Peru and the Seychelles Islands and testing for resistance to lethal yellowing disease. CPL 17. Rome: Food and Agriculture Organisation

Genetic markers for coconut palms

G.R. ASHBURNER
Agriculture Victoria, Institute for Horticultural Development, Private Bag 15, South Eastern Mail Centre Vic. 3176, Australia
Present address: Centro de Investigación Científica de Yucatán, Apartado Postal 87, Cordemex, 97310 Mérida Yuc., México

1. Introduction

Any measurable character which is inherited in a simple Mendelian fashion can be used to identify an individual, a group of individuals or a position on a chromosome, and thus can be used as a genetic marker. Therefore, genetic markers usually result from the action of one or two major gene loci, and differ from polygenic characters such as yield. These markers may be obvious morphological characters such as spicata inflorescences, or may be at the DNA sequence level which can only be detected with complicated biochemical assays. Genetic markers may be either monomorphic or polymorphic, but generally speaking, must be polymorphic to be useful in plant breeding. The utility of the genetic marker depends upon the purpose it is to be used for and how practical it is to assess. This paper provides an introduction to both the current and the potential use of genetic markers in coconut breeding and genetics.

2. Common genetic markers

2.1 Morphological markers

Morphological markers are morphological characters with an established inheritance. They are simply measured because they are assessed visually. However, the number of useful morphological markers is limited because their phenotypic expression is often affected by environmental conditions, polygenic inheritance, and partial or complete dominance. Still, the field of genetics is based

C. Oropeza et al. (eds.), Lethal Yellowing: Research and Practical Aspects, 173–186.

upon the pioneering work of Mendel [23] who used just such morphological markers.

Very few morphological genetic markers have been characterised for coconut palms, possibly because the study of coconut genetics is very long-term and therefore has not attracted major resources. In addition, strong canalising natural selection has resulted in seemingly limited diversity. Documented examples of morphological markers in coconut palms include: spicata inflorescence which is conferred by a single dominant gene [28], makapuno fruit conferred by a single recessive gene [39] and fruit/shoot/petiole colour in dwarfs which is thought to be due to the action of two loci with two alleles each [5].

2.2 Disease resistance markers

Expression of disease resistance markers is shown by resistance or tolerance to diseases. Such markers are widely used in plant breeding since they are usually traits of great economic importance. As with morphological markers, the expression of disease resistance/tolerance may be affected by environmental factors and assessment may be time-consuming and costly. The number of markers available is obviously limited to the number of diseases affecting the plant species.

Resistance/tolerance have been described for lethal yellowing in Jamaica [17] and coconut foliar decay virus (CFDV) in Vanuatu [9]. There are reports of resistance to *Phytophthora palmivora* (Butler) Butler *sensu lato*; *P. katsurae* Ko & Chang and *Bipolaris incurvata* (Ch. Bernard) Alcorn. (Syn. *Drechslera incurvata*; *Helminthosporium incurvatum*) in various coconut populations [13], but they require further evaluation.

2.3 Isozyme markers

Isozymes are specific enzymes with multiple forms. The different forms of enzymes arise due to variation in loci or variation in alleles at these loci. Different loci and differing alleles are detected by differences in enzyme migration during electrophoresis. Allelic variations arise due to differing amino acid complements resulting from gene mutation, conferring different net charges and structures to the proteins. Since isozymes are primary products of DNA translation, they are less affected by environmental influences than are gross morphological characters, and their inheritance is relatively simple to deduce. Isozyme analysis is widely used in plant breeding and genetics and the methods are well characterised [36].

Isozyme analysis has a number of advantages over other techniques for defining genetic markers [8]. The technique is easy, inexpensive and only small samples are required. Samples can also be assessed at a laboratory remote to the sampling area. The markers are expressed in a co-dominant manner making heterozygotes simple to observe. However, there are a number of limitations [7]: there may be post-translational modifications, only one quarter of mutations lead to protein differences, and they are restricted to only one class of proteins, meaning that there is a limited number of possible markers. Since isozyme markers are proteins, they spoil easily and the samples must be treated accordingly.

Isozyme analysis has been widely used in the family Arecaceae, for example with *Elaeis guineensis* [2, 14], *E. oleifera* [15], *Phoenix dactylifera* [3, 38], *Washingtonia filifera* [22] and *Astrocaryum mexicanum* [10]. Isozyme analysis has also been used for coconut palms [4] (Moran pers. comm., Cardeña pers. comm., Ashburner unpubl.). Fifty isozyme loci have been described for coconut pollen using 26 enzyme systems (Table 1). Various tissue has been evaluated for enzyme activity with pollen and embryos showing the best results. Pollen is the preferred tissue because it is simple to transport and store, and does not suffer from developmental variation in the isozyme profile.

Table 1. Isozyme systems described for coconut pollen [4] (Moran pers. comm.)

Enzyme	No of loci	Enzyme	No of loci
Acid phosphatase	1	Isocitrate dehydrogenase	3
Aconitrate hydratase	2	Leucine amino peptidase	2
Alanine amino peptidase	2	Malate dehydrogenase	3
Alcohol dehydrogenase	2	Malate reductase	1
Aspartate amino transferase	3	Mannosephosphate isomerase	1
Diaphorase	2	Peroxidase	1
Esterase	2	Phosphagluconate dehydrogenase	2
Fructose-1,6-isomerase	1	Phosphoenol pyruvate carboxylase	1
Glucose-6-phosphate dehydrogenase	1	Phosphoglucomutase	5
Glucosephosphate isomerase	3	Shikimate dehydrogenase	1
Glutamate dehydrogenase	1	Sorbitol dehydrogenase	1
Glutamate oxaloacetate	4	Superoxide dismutase	1
Glycerate dehydrogenase	3		

2.4 *RFLP markers*

Restriction Fragment Length Polymorphism (RFLP) analysis is a DNA-based technique. RFLP analysis measures the size of different fragments of the genome after it has been digested with restriction endonucleases which cleave DNA at specific sites (restriction sites). Polymorphisms are generated when there is a deletion or addition between two restriction sites, or when there is a mutation at the restriction site resulting in a change in fragment length. The polymorphisms are usually visualised using gel electrophoresis, blotting and DNA hybridisation with radioactive probes. Many of these markers will be monomorphic. RFLP analysis is being used in the genetic improvement of many different crop species [30] and in plant systematics [35].

RFLP analysis elucidates the genotype of the organism, so there are no environmental or developmental effects on expression. Due to the large size of plant genomes and the number of potential restriction sites, there is a large number of potential RFLP markers. RFLP markers are useful because they are not restricted to coding regions of the genome, they are expressed in a co-dominant manner and may have multiple alleles. Remote collection of material is possible because the DNA is relatively stable. However, the method is both laborious and expensive.

Within the Arecaceae, RFLP markers are currently used with *Elaeis guineensis* [19] and have been used in systematic studies of the family [43]. RFLP markers are currently being developed for *Cocos nucifera* by using low-copy number genomic clones and various cloned genes [32] (Rohde pers. comm.).

2.5 *RAPD markers*

Random Amplified Polymorphic DNA (RAPD) analysis is a DNA technique based on Polymerase Chain Reaction (PCR) technology. PCR makes it possible to generate large quantities of DNA between any two specific priming sites from minute quantities of template DNA [11]. RAPD utilises this technique by amplifying DNA from random regions of the genome by using random primers (10-mers). Different random primers amplify different random regions. The results are visualised by gel electrophoresis and scored by comparing fragment lengths between individuals. Polymorphisms are created when there are mutations in the priming site or when there are insertions or deletions between the priming sites.

Each locus has only two possible alleles: either presence or absence of the fragment, and therefore the characters are inherited in a dominant manner. For a review of the RAPD technique and its applications see Waugh and Powell [40] and Yu *et al.* [44].

Since RAPD is a DNA based marker technique, it shares the same advantages as RFLP markers. However, it is simpler, quicker, cheaper and visualises more independent loci in an assay than RFLP. A disadvantage is that RAPD markers are dominant and heterozygotes cannot be distinguished. RAPD also suffers from being very sensitive to assay conditions [24] meaning that the markers are not readily transferable between laboratories.

RAPD is still a relatively new technique and little use has been made of it in the Arecaceae. The technique is readily applicable for creation of molecular markers for coconut palms [1]. DNA is easily extracted from leaves using a method based upon Rogers and Bendich [33]. RAPD fragments are generated in a 25 μl reaction containing 67 mM Tris-HCl pH 8.8, 16.6 mM $(NH_4)_2SO_4$, 0.45% Triton X-100, 0.02% gelatine, 4 mM $MgCl_2$, 200 μM each of dATP, dCTP, dGTP, dTTP, 0.25 μM 10-mer (Operon Technologies), approximately 0.25 μg genomic DNA and 1U *Taq* DNA polymerase, all overlaid with 1 drop paraffin oil. Thermal cycling is performed in a DNA Thermal Cycler programmed for 45 cycles of 60 s at 94°C, 60 s at 36°C and 120 s at 72°C. Amplification products are visualised by electrophoresis on a 1.5% agarose gel in TAE buffer at 75 V for 4.5 h.

2.6 VNTR markers

Variable Number Tandem Repeat (VNTR) analysis may be regarded as a form of RFLP analysis. VNTRs consist of stretches of tandemly repeating nucleotide sequences which occur frequently throughout the genome. They almost always show extensive polymorphism due to site-specific length variation caused by variability in the number of repeat units. The length variations of these sequences are visualised by RFLP techniques [*e.g.* 29] where many loci are visualised and inheritance is difficult to determine. Since the number of potential loci and alleles is great, the technique is likely to generate an abundance of genetic markers. VNTR technology has been used with few plant species to date.

2.7 Sequence Tag Sites

The DNA sequence of RFLP, RAPD and VNTR markers can be determined using molecular biological techniques. With this information, primers can be designed that will amplify the given polymorphisms using PCR technology. These primers are known as Sequence Tag Sites (STS) [31]. STS can overcome many of the practical problems involved with routine DNA marker assays. RFLP markers become fast and inexpensive to visualise by using PCR technology. RAPD markers become more reliable and portable because longer and more specific primers are used during PCR reactions. VNTR markers are expressed as single loci instead of multiple loci making their inheritance and allelic variation simple to determine [26]. Since STS markers are time-consuming and costly to develop, they should only be used to create markers of proven utility. Use of STS markers will increase as the costs of PCR and DNA sequencing falls.

3. Use of genetic markers in coconut improvement

3.1 Genotype characterisation

By selecting markers specific to parental types, their use ensures purity of progeny by screening them for inheritance of those markers. For centuries, the Polynesians of Samoa are thought to have been practising this procedure with fruit/shoot colour markers to ensure purity of the 'Niu Vai' and 'Niu 'Afa' ecotypes of coconuts [12]. It is probable that a similar quality control measure was used in some regions to ensure the genetic purity of dwarf ecotypes [18]. Fruit/shoot colour markers are currently used to verify the hybrid status of intraspecific crosses between dwarf and tall ecotypes [42]. These markers are used to indicate the selfs of crosses between red or yellow dwarf mothers and tall pollen donors.

Although colour markers are currently the method of choice for ensuring purity of the crosses, isozymes or RAPD could be used if no colour markers were available. Given that isozyme diversity between populations is relatively low in coconuts (Moran pers. comm.), RAPD is probably a more useful technique. RFLP and VNTR analysis have potential if used to create STS markers.

Genetic markers may also be used to characterise germplasm. In particular, the frequencies of the markers may be used to estimate germplasm diversity, population structure, and to construct phylogenies. This information is necessary when formulating germplasm collection, conservation and utilisation strategies. It is particularly important in coconut palms because they have recalcitrant seeds and therefore germplasm must be maintained in costly field plots. Genetic markers may also be used to set priorities in crossing programs if genetic distances between populations could be used as an indication of potential heterosis.

Frequencies of markers are the simplest descriptors of diversity. These, in turn, may be used to calculate the heterozygousity of the population if the markers are co-dominant or if the conditions of Hardy-Weinberg equilibrium are met. For coconut germplasm analysis, it is tenuous to assume Hardy-Weinberg equilibrium. When data are available for several markers, it is possible to determine the genetic distance between different populations. These distances may, in turn, be used to construct a distance matrix between populations from which dendrograms may be constructed through cluster analysis. For further details refer to Nei [27] and Weir [41]. Characters of unknown inheritance such as foliar polyphenols [20] may be used to the same effect by using numerical taxonomy techniques [34].

Coconut germplasm diversity has been characterised by using disease resistance, isozyme and RAPD markers. Isozyme analysis has revealed limited genetic differentiation between coconut populations. [4] (Moran pers. comm.). A study of Melanesian coconut populations using 15 enzyme systems, revealed that 46.9% of the loci were polymorphic and there was an average of 1.59 alleles per loci (Moran pers. comm.). These results are comparable to other monocots [16]. However, many of the polymorphisms in this study were due to rare random alleles, other loci showed low levels of variability in all populations, and no specific population markers were found. RAPD markers have also been used to characterise south Pacific germplasm diversity [1]. In that study, it was evident that there was a high degree of diversity both within- and between-populations. RAPD markers unique to a population were rare and most markers occured in all populations. However, a considerable number of unique RAPD markers separated Pacific and Indian Ocean coconut types. Lethal yellowing resistance/tolerance has also been used to discriminate between coconut populations and as an index of domesticated-type introgression into the wild-type genepool [18].

3.2 Breeding system characterisation

Genetic markers are commonly used to characterise breeding systems of plants. Knowledge of the breeding system is required to explain population structures and to ensure that correct conservation strategies are established. In the simplest approach, estimates of the proportion of offspring produced by outcrossing are based on the proportion of heterozygous offspring carrying one allele which could not be of maternal origin. Various models have been used to estimate outcrossing depending on whether the maternal genotype is known or whether multiple loci are involved [6].

Recessive fruit colour markers have been used to estimate minimum outcrossing rates in dwarf ecotypes [5, 42] and one population of talls in Papua New Guinea [1]. Isozymes are commonly used in these studies because they are cheap and simple. However, isozymes may not be as useful with coconut palms because they appear to lack sufficient diversity. RAPD analysis is precluded from these studies because they are dominant markers and RFLP is too expensive and laborious. STS has the potential to finally determine the breeding system of tall coconut ecotypes.

3.3 Genetic mapping and trait linkage

Genetic markers are now contributing to plant breeding and genetics by being used to create genetic maps. A genetic map is a visual representation of the physical location of loci or genetic markers on the chromosomes. DNA markers are placed into the context of the map by their linkage with one another. Genetic markers which segregate together in a segregating population (*e.g.* F_2 population from a diverse cross) are linked and therefore are located on the same chromosome. If linkage on the chromosome is close, they segregate together, if it is further apart, they mostly segregate together but varies due to crossing-over events. Once a genetic map has been established with genetic markers, other traits may be included. Then genetic markers which are closely linked to a particular trait may be used to select individuals that will exhibit that trait. This is known as "gene tagging". It is particularly useful when the tagged character is costly or laborious to assess. Such traits may be loci for quantitative traits and are so termed Quantitative Trait Loci (QTL). Many loci are involved in the trait and the expression of these loci are confused by other contributing loci. Therefore, if there are markers closely linked to the desirable alleles of these QTL, it is possible to select for individuals expressing those alleles [37].

The choice of the crosses to be made in developing genetic maps and tagging genes depends on the marker system used and the inheritance of the traits being targeted. If co-dominant markers such as RFLP are used, analysis usually involves the F_2 population from a cross between two diverse individuals [21]. At least one of these individuals should be expressing traits of interest. If QTL are being targeted, then different crosses are needed to analyse each locus involved in the quantitative trait after it has been identified in a wide cross. If dominant markers such as RAPD are being used, the same protocol may be followed but will be laborious because heterozygotes cannot be easily determined. It is also uncertain that a particular RAPD locus will be amplified in different crosses, therefore analysis is restricted to the original cross from which the segregating population was constructed. A number of different strategies have been devised to exploit the speed of RAPD marker generation that compensate for its inherent problems [40]. If single loci are being targeted, comparison of Near-Isogenic Lines (NIL) will reveal polymorphisms that are likely to be closely linked to the locus of interest. Similarly, the use of bulked DNA samples from the homozygotes of a trait in a F_2 population are likely to express polymorphisms that are closely linked to the trait of interest. When looking at quantitative traits, bulked DNA samples from extreme members of the trait distribution can show polymorphisms that are linked to quantitative loci. The efficiency of this technique is greatly improved if Recombinant Inbred Lines (RIL) or doubled haploids are used to create homozygous traits and markers.

For coconut palms, possibilities include tagging of major genes such as resistance/tolerance to lethal yellowing, CFDV and other pests and diseases, once their inheritance has been determined. Marker-assisted selection is also possible using QTL tags for traits such as drought resistance, oil content and yield whose components have a large additive effect [24]. To this end, a genome mapping initiative has been established at the Max Planck Institute, Germany, which will identify both RFLP and RAPD markers (Rohde pers. comm.).

Morphological and isozyme markers have little potential in the creation of coconut genetic maps since they have limited polymorphisms. It remains to be seen whether RFLP will supply more variation, but very early results indicate some useful variation (Ashburner unpubl.). The RAPD technique has revealed variation within- and between-populations in the south Pacific and great variation between Pacific and Indian Ocean populations [1]. VNTR technology promises to provide even more polymorphic markers to add to the genetic map.

The creation of genetic markers linked to important traits is long-term because of the long generation time of coconuts and the infancy of breeding efforts. RFLP markers appear to have the greatest potential at present but populations segregating for useful traits must be created. RAPD analysis may be used in specific cases such as developing markers for lethal yellowing resistance/tolerance or CFDV tolerance, but F_2 and F_2 progeny tests must be initiated. Creation of NILs are precluded because of the time required. Use of RAPD for QTL-tagging has potential, but will also be relatively slow because RILs and doubled-haploids do not exist in coconut breeding. RAPD analysis of segregating populations resulting from domesticated-type introgression into wild-type gene-pools is one novel option that may be explored.

4. Conclusions

Genetic markers are set to revolutionise coconut breeding through the use of marker assisted selection and germplasm characterisation. However, a number of issues must be resolved first. There must be co-operation between breeders, geneticists, physiologists, agronomists, pathologists and entomologists to uncover traits that are amenable to gene-tagging. Mapping populations must then be established which incorporate these traits. A co-ordinated approach is also needed to create the markers, dedicated laboratories must work with breeders to create markers from the relevant populations and then provide a simple and cheap product that can be used by those breeders. PCR-based STS markers are well suited to this role. How markers are to be used in breeding programs must also be considered. Will marker-assisted selection be used to incorporate desirable traits into parental lines in a F_1 breeding program, and if so, will they also be used to enhance the complementarity of those parental lines for both QTL and major gene loci? Or will a new breeding protocol be developed? The challenge has been issued. Coconuts, welcome to the 21st century.

Acknowledgements

This work was supported by the Australian Centre for International Agricultural Research in collaboration with the Cocoa and Coconut Research Institute, Papua New Guinea. Travel assistance was provided by Centro de Investigación Científica de Yucatán, México. The following personal communications are acknowledged: R. Cardeña, CICY, México, G.F. Moran, CSIRO Division of Forestry and Forest Products, Australia, and W. Rohde, Max Planck Institute, Germany.

References

1. Ashburner GR (1994) Characterisation, collection and conservation of *Cocos nucifera* L. in the south Pacific. PhD thesis, University of Melbourne
2. Baudouin L (1992) Utilisation des marqueurs moléculaires pour l'amélioration du palmier à huile I.- Marqueurs protéiques. Oléagineux 47:681-691
3. Bennaceur M, Lanaud C, Chevallier MH and Bounaga N (1991) Genetic diversity of the date palm (*Phoenix dactylifera* L.) from Algeria revealed by enzyme markers. Plant Breed 107:56-69
4. Benoît H and Ghesquière M (1984) Electrophorèse, compte-rendu cocotier. IV. Déterminisme génétique. Internal Report, Paris: IRHO-CIRAD
5. Bourdeix R (1988) Etude du déterminisme génétique de la couleur du germe chez le cocotier nain. Oléagineux 43:371-374
6. Brown AHD (1989) Genetic characterization of plant mating systems. In: AHD Brown, MT Clegg, AL Kahler and BS Weir, eds. Plant Population Genetics, Breeding, and Genetic Resources, pp 145-162 Sunderland: Sinauer Associates
7. Brown AHD and Moran GF (1981) Isozymes and the genetic resources of forest trees. In: MT Conkle, ed. Isozymes of North American Forest Trees and Forest Insects, pp 1-10. Berkeley: USDA
8. Brown AHD and Weir BS (1983). Measuring genetic variability in plant populations. In: SD Tanksley and TJ Orton, eds. Isozymes in Plant Genetics and Breeding, Part A, pp 219-239. Amsterdam: Elsevier
9. Calvez CH, Renard JL and Marty G (1980) La tolérance du cocotier hybride Local x Rennell à la maladie des Nouvelles-Hébrides. Oléagineux 35:443-451
10. Eguiarte LE, Perez-Nasser N and Piñero D (1992) Genetic structure, outcrossing rate and heterosis in *Astrocaryum mexicanum* (tropical palm): implications for evolution and conservation. Hered 69:217-228
11. Erlich HA, ed. (1989) PCR technology: principles and applications for DNA amplification. New York: Stockton Press
12. Foale MA (1987) Coconut germplasm in the south Pacific Islands. ACIAR Technical Report No 4. Canberra: Australian Centre for International Agricultural Research
13. Foale MA (1992) Coconut diversity- present knowledge and future research needs. In: Coconut Genetic Resources. Papers of an IBPGR Workshop, Cipanas, Indonesia, 8-11 October, 1991. International Crop Network Series No. 8, pp 46-58. Rome: International Board for Plant Genetic Resources
14. Ghesquière M (1984) Polymorphisme enzymatique chez le palmier à huile (*Elaeis guineensis* Jacq.). I. Contrôle génétique de neuf systèmes enzymatiques. Oléagineux 36:475-486
15. Ghesquière M, Barcelos E, Santos MM and Amblard P (1987) Polymorphisme enzymatique chez *Elaeis oleifera* HBK (*E. melanococca*). Analyse des populations du Bassin amazonien. Oléagineux 42:143-150
16. Hamrick JL and Godt MJ (1989) Allozyme diversity in plant species. In: AHD Brown, MT Clegg, AL Kahler and BS Weir, eds. Plant Population Genetics, Breeding, and Genetic Resources, pp 43-63. Sunderland: Sinauer Associates
17. Harries HC (1970) The Malayan Dwarf supersedes the Jamaica Tall coconut. 1. Reputation and performance. Oléagineux 25:527-531
18. Harries HC (1978) The evolution, dissemination and classification of *Cocos nucifera* L.. Bot Rev 44:265-317
19. Jack PL and Mayes S (1993). Use of molecular markers for oil palm breeding II.-Use of DNA markers (RFLPs). Oléagineux 48:1-8
20. Jay M, Bourdeix R, Potier F and Sanslaville C (1989) Premiers resultats de l'etude du polymorphisme des polyphenols foliaires du cocotier. Oléagineux 44:151-161
21. Landry BS (1993) DNA mapping in plants. In: BR Glick and JE Thompson, eds. Methods in Plant Molecular Biology and Biotechnology, pp 269-285. Boca Raton: CRC Press

22. McClenaghan LR and Beauchamp AC (1986) Low genic differentiation among isolated populations of the Californian fan palm (*Washingtonia filifera*). Evol 40:315-322
23. Mendel G (1866) Versuche über pflanzen-hybriden. *Verhandlungen des naturforschenden Vereines in Brunn* 4:3-47. [Reprinted in: Peters JA (1959) Classic Papers in Genetics. Englewood Cliffs: Prentice Hall]
24. Meunier J, Sangaré A, Saint JP Le and Bonnot F (1984) Analyse génétique des caractères du rendement chez quelques hybrides de cocotier *Cocos nucifera* L.. Oléagineux 39:581-586
25. Meunier JR and Grimont PAD (1993) Factors affecting reproducibility of random amplified polymorphic DNA fingerprinting. Res Microbiol 144:373-379
26. Morgante M and Olivieri AM (1993) PCR-amplified microsatellites as markers in plant genetics. Plant J 3:175-182
27. Nei M (1987) Molecular evolutionary genetics. New York: Columbia University Press
28. Ninan CA and Satybalan K (1963) Cytogenetic studies in the genus *Cocos*. II. Some observations on the 'spicata' character in coconuts (*Cocos nucifera* L.). Indian Coconut J 16:109-114
29. Nybom H and Schaal BA (1990) DNA "fingerprints" applied to paternity analysis in apples (*Malus* X *domestica*). Theor Appl Genet 78:847-851
30. O'Brien SJ (1990) Genetic maps: locus maps of complex genomes. New York: Cold Spring Harbour Laboratory Press
31. Olson M, Hood L, Cantor C and Botstein D (1989) A common language for physical mapping of the human genome. Science 245:1434
32. Rohde W, Salamini F, Ashburner GR and Randles JW (1992) An *Eco*RI repetitive sequence family of the coconut palm *Cocos nucifera* L. shows sequence homology to *copia*-like elements. J Genet Breed 46:391-394
33. Rogers SO and Bendich AJ (1985) Extraction of DNA from milligram amounts of fresh, herbarium and mummified plant tissue. Plant Mol Biol 5:69-76
34. Sneath PHA and Sokal RR (1973) Numerical Taxonomy. San Fransisco: Freeman
35. Soltis PS, Soltis DE and Doyle JJ, eds (1992) Molecular Systematics of Plants. New York: Chapman and Hall
36. Tanksley SD and Orton TJ, Eds (1983) Isozymes in Plant Genetics and Breeding, Part A. Amsterdam: Elsevier
37. Tanksley SD, Young ND, Paterson AH and Bonierbale MW (1989) RFLP mapping in plant breeding: new tools for an old science. Bio/Tech 7:257-264
38. Torres AM and Tisserat B (1980) Leaf isozymes as genetic markers in date palms. Am J Bot 67:162-167
39. Torres J (1937) Some notes on makapuno coconut and its inheritance. Philipp J Agric 8:27-37
40. Waugh R and Powell W (1992) Using RAPD markers for crop improvement. TIB 10:186-191
41. Weir BS (1990) Genetic Data Analysis. Sunderland: Sinauer Associates
42. Whitehead RA, Thompson BE and Williams LV (1966) A genetic marker of use in coconut seed production. Oléagineux 21:15-16
43. Wilson MA, Gaut B and Clegg MT (1990) Chloroplast DNA evolves slowly in the palm family (Arecaceae). Mol Biol Evol 7:303-314
44. Yu KF, Van Deynze A and Pauls KP (1993). Random Amplified Polymorphic DNA (RAPD) analysis. In: BR Glick and JE Thompson, eds. Methods in Plant Molecular Biology and Biotechnology, pp 287-301. Boca Raton: CRC Press

Glossary

Some definitions are derived from Waugh and Powell [40].

Alleles- two or more genes that occupy the same relative position (locus) on a pair of homologous chromosomes and undergo pairing at meiosis, produce different effects on the same set of developmental processes and can mutate one to another.

Backcross- the generation derived from crossing a hybrid with one of its original parents, usually the female.

Co-dominant- expression of both alleles at a given locus.

Dominant- expression of only one allele at a given locus, the other allele is recessive.

Doubled haploid- a sporophyte that has undergone chromosome doubling resulting in complete homozygousity.

F_1- the progeny derived from crossing plants or animals of the parental generation (P_1).

F_2- the progeny derived from crossing the members of the F_1 among themselves (Selfing).

Genetic distance- a measure of population divergence, an absence of difference between populations is zero while populations with no common alleles has a genetic distance of one.

Heterozygote- having two different alleles at the corresponding loci of a pair of chromosomes.

Homozygote- have identical genes at the corresponding loci of a pair of chromosomes.

Locus- the position occupied in homologous chromosomes by a particular gene or one of its alleles.

Monomorphic- exhibiting one form (or allele).

Near Isogenic Line- a genotype differing from another only in a particular gene of interest. Produced by repeated backcrossing with selection for that character.

Polymorphic- exhibiting two or more forms (or alleles).

Polygenic- characters controlled by more that one gene locus and are influenced by genetic and environmental factors. They tend to exhibit a continuous range of phenotypes.

Primer- a short sequence of artificially produced DNA that binds to a complementary region of DNA and primes DNA polymerisation during a PCR event.

Progeny testing- a method for assessing the genotype of an individual by a study of its offspring.

Quantitative Trait Loci- gene loci that influence polygenic characters.

Recombinant Inbred Line- a genotype created by repeated self-pollination.

Segregation- the separation and distribution of alleles to different cells (and therefore progeny) following meiosis.

Production and advantages of coconut hybrids

BASIL O. BEEN
Coconut Industry Board, 18 Waterloo Rd, Kingston 10, Jamaica

1. Introduction

Maize (*Zea mays*) breeders have made "hybrid" a household word, and, given the success of hybrid maize, it was natural for breeders of other crops to concentrate on hybridisation as an improvement method. The commercial production of coconut F_1 hybrids has now become an important aspect of coconut development schemes in producing countries world-wide. Details of hybrid production vary with personal preference and available equipment but, in general, the steps involved are selection of parent trees, emasculation, isolation, pollination and harvesting of seed.

2. Floral biology of coconut palms

A knowledge of the floral biology of the coconut is essential before controlled pollination is undertaken (Ashburner, this volume). The onset of flowering in coconut palms varies according to ecotype and environment, after which a new inflorescence opens approximately every month. The inflorescence consists of a central axis with lateral branches which bear numerous male flowers terminally and one or more female flowers at the base. The male flowers mature first, beginning terminally on each branch and proceeding towards the base. The male flowers fall after the pollen has been shed. The male phase begins with the opening of the first male flower and ends with the shedding of the last male flower. The female phase is the interval between the first female flower becoming receptive and the last becoming post-receptive. The male phase lasts for approximately three weeks; the female phase about five to seven days and usually begins two to three weeks after the spathe opens. The lengths of the phases for any given ecotype are influenced by locality and season. In most tall ecotypes, male

C. Oropeza et al. (eds.), Lethal Yellowing: Research and Practical Aspects, 187–194.

and female phases do not overlap in the inflorescence; all male flowers on an inflorescence are shed before the female flowers become receptive. In most dwarf ecotypes, there is overlapping of the phases. The lengths of male and female phases of three ecotypes widely cultivated in Jamaica are presented in Table 1.

Table 1. Lengths of male and female phases of three 'varieties' of coconut grown in Jamaica.

Ecotype	Mean length of male phase (days)	Mean length of female phase (days)	Interval between male and female phases (days)
Jamaica Tall	22.1	5.9	2.9
Panama Tall	23.1	4.7	2.2
Malayan Dwarf	23.4	5.6	-

Coconut fruits mature between 11 and 16 months after pollination, and is dependant upon the ecotype and environment.

The flowering pattern of coconut palms makes it possible to maintain a continuous controlled pollination program. The annual fruit production is generally less than one hundred per palm, each fruit contains only one seed, and it is usually necessary to climb the palm to carry out emasculations and pollinations.

3. Hybridisation techniques

The essential elements of hybridisation are emasculation, isolation and pollination.

3.1 Emasculation

All male flowers are removed by cutting each rachilla of the inflorescence about 2-3 cm above the female flowers and picking off those male flowers near the base of the female flowers. To minimise the risk of undesirable shedding of pollen, it is advisable to carry out emasculations by "pre-bursting" the spathes about a day before they would normally open.

3.2 Isolation

Exclusion of unwanted pollen from emasculated inflorescences may be achieved by:

a) use of a physical barrier, bags or polycaps, to enclose the entire inflorescence or individual flowers, and

b) use of space isolation.

Pollination bags should be large enough to comfortably enclose the emasculated inflorescence. Several materials are used to make bags, but whatever material is used, it should be weather proof and permeable to water vapour; polythene and rubber are unsuitable. Bags must be tied securely in place with the seal between bag and inflorescence stalk being made pollen- and insect-proof with cotton or Styrofoam impregnated with insecticide. The use of polycaps made of soft polyurethane foam to isolate individual flowers has been suggested by Harries [1].

If large quantities of F_1 hybrids are to be produced, isolation of individual inflorescences or flowers would not only be impractical, but also very expensive. Therefore, in the commercial production of hybrids, isolation is best achieved by planting mother palms in an area where no undesirable coconuts are grown, or by surrounding them with a screen of foliage. This forest barrier should be about 300 m wide [2, 4].

3.3 Pollination

When female flowers become receptive, pollen is applied by swab, brush, puffer-bottle or blower. Where mother palms are interplanted with pollen parents, pollination of emasculated inflorescences occurs naturally. The female flower is receptive when the stigma is reflexed, moist and white. Browning of the stigma indicates post-receptivity.

4. Commercial production of F_1 hybrids in Jamaica

In Jamaica, F_1 hybrids have been produced commercially by mass controlled pollination [1] since 1971 when it became clear that the two-parent hybrid seed-gardens, which had been established earlier, were too inflexible for large-scale hybrid seed production.

4.1 Selection of parents

The main F_1 hybrid being produced in Jamaica is the 'Maypan'. In it are combined the lethal yellowing resistance of the 'Malayan Dwarf', the female parent, and the hardiness of the 'Panama Tall', the pollen parent.

To be able to differentiate between genuine hybrid seedlings and those resulting from unwanted self-pollination, use is made of colour inheritance when choosing parents. By using yellow- and red-fruited 'Malayan Dwarfs' as female parents, hybrids can be selected in the nursery on the basis of petiole colour, and apparent selfs can be eliminated. When 'Malayan Dwarf' x tall hybrids are produced, the seedlings usually have the petiole colours of the tall (Table 2).

Table 2. Colour inheritance (simplified) of dwarf x tall coconut F_1 hybrids.

Female parent	Male parent				
	Yellow dwarf	Red dwarf	Green dwarf	Green tall	Bronze tall
Yellow dwarf	yellow	red	green	green	bronze
Red dwarf	red	red	green	green	bronze
Green dwarf	green	green	green	green	bronze

4.2 Emasculation

Emasculation is carried out as described. On the day before the inflorescence would normally open, a worker climbs the tree, cuts the tip of the spathe with a pair of secateurs and then splits it vertically into two or three strips that are cut off at the base. The male flowers are then removed. Great care must be exercised in effecting emasculation because the inflorescence tissue at this stage is tender and the immature female flowers can be easily damaged. If there is an over-abundance of female flowers, some are removed. Having completed the emasculation, the worker writes the date on the petiole of the subtending leaf and enters the palm number on his emasculation/pollination record sheet. Male flowers removed by emasculation can be disposed of in a variety of ways, *e.g.* placing in pits, bottomless drums or under dry fallen fronds.

4.3 Isolation

Isolation is achieved by establishing the seed garden in an area where no other coconuts are grown and every palm is numbered for purposes of control.

4.4 Pollen handling

4.4.1 Collection

When approximately 25% of the male flowers have fallen from the inflorescence, spikelets carrying the remaining male flowers are cut from the desired palm. We have found that at this stage, it is possible to obtain a satisfactory yield of viable pollen.

The male flowers are then stripped from the spikelets, cracked using counter-rotating rollers and dried in a fluid-bed drier at 40-50°C for 3-4 h. After drying, flowers are sieved using a mechanical shaker. Using this method, a batch of 3 kg of fresh flowers yields about 60 g of pollen with 40-50% viability. Small (less than 3 kg) quantities of flowers are sometimes dried in an oven at 40°C for 24-40 h. Flower drying can also be carried out in specially-designed drying-rooms.

Pollen is dust-like and is therefore subject to contamination, especially when dealing with several different lots. Therefore, it is necessary to wash hands, laboratory benches, storage containers *etc*., between different lots.

4.4.2 Storage

Pollen is living and may be killed by sub-optimal storage. Fortunately, coconut pollen, unlike some others, has a relatively long life. Having been extracted, the pollen (if not used immediately) is stored in closed jars for a number of weeks in a refrigerator (5°C), or for a few months in a deep freeze (-20°C). Vacuum freeze-drying is used for long-term storage.

4.4.3 Testing

Before the pollen is used, its viability is laboratory tested. A culture medium containing sugar, gelatine and distilled water in the ratio 2:7:20 is used. These three ingredients are placed in a container, mixed and placed in a pressure-cooker long enough to attain 15 lb pressure. The mixture is then poured into petri-dishes (to form a thin layer) which are immediately covered. Once the medium has cooled, pollen is dusted lightly on the surface of the medium and incubated at 30°C for 1-5 h. Germinated pollen grains can be counted under a microscope at low magnification. Only grains with pollen tubes longer than the diameter of the grain are considered to have germinated. Batches of pollen with viability lower than 30% are not used in hybrid seed-gardens.

4.4.4 Diluent

Before it is used in the field, pollen is diluted by mixing talc or chalk in the ratio of 1 part pollen to 3 parts diluent.

4.5 Pollination

Pollen is applied to receptive female flowers by workers using polyurethane swabs. Every day, each worker is given enough pollen/talc mixture to last for that day, with unused pollen being discarded at the day's end. Records are kept of each pollination. When there is to be a change of pollen source, there is one week's break in pollination. Inflorescences pollinated with the new pollen source are tagged to avoid confusion at harvest.

4.6 Fruit harvesting and handling

During their long development period, fruit may be affected by pests and diseases which can lead to premature nut-fall. Prophylactic spraying with Ridomil helps to reduce nut-fall associated with *Phytophthora* and use of rodenticides minimises rat-damage.

Fruit are reaped when they begin to lose their original colour and become grey-brown. Defective fruit are discarded and those remaining are set in labelled nursery beds where germination then proceeds. After germination, hybrids can be recognised by their petiole colour.

4.7 *Quality control*

The success of controlled pollination depends upon a number of factors such as: efficiency of the work-force, viability of the pollen, physiological state of the palms, the environment and difficulty in climbing and carrying out the necessary operations.

We have determined that one worker can efficiently emasculate and pollinate a block of 400 mother palms. An effective monitoring system must be in place at all times as the efficiency of the emasculation/pollination operations will not be determined until over a year after the operation.

5. Advantages of hybrids

The object of crossing plants is to combine, in a single variety, the desirable characters of two or more lines, varieties or species. To be considered for large-scale use, hybrids must possess hybrid vigour or some combination of traits that make them more valuable than their parents, and be capable of being mass-produced.

Crosses between different varieties occur naturally but the earliest work on hybrid production was done in Fiji [3]. Since then, a large number of F_1 hybrids have been produced in many countries. The breeding programs generally conform to the following pattern: assessment of local and introduced germplasm, hybridisation of promising varieties, testing of F_1 hybrids and large-scale production of those selected. Hybridisation is considered the quickest means of improving yields and quality of coconut palms.

Until recently, emphasis was placed on dwarf x tall hybrids to combine the precocity, dwarfness, high nut production (and disease resistance) of the dwarf, with the large nut size, good copra quality and robustness of the tall. Tall x tall hybrids are also being seriously considered in some countries. Over the past thirty years, a number of F_1 hybrids that demonstrate precocity, disease resistance and high yield potential, have been used with great success in development and rehabilitation programs; *e.g.* 'PB-111' and 'PB-121' developed by IRHO (now CIRAD-CP), and 'Maypan' developed by the Coconut Industry Board, Jamaica.

In many crops, hybridisation is usually used in the initial stages of a breeding project to generate a variable population from which they can subsequently select plants exhibiting the best combination of traits. The usual practice is to start with two lines that possess the desired traits, cross these, plant the F_1 so that they cross among themselves or self, plant and select among the F_2 hybrids, plant and select the F_3 hybrids and so on. This approach of recurrent selection has not appealed to coconut breeders because of the long generation lengths and large sizes of populations.

Acknowledgements

This paper was published with permission from the Coconut Industry Board, Jamaica.

References

1. Harries HC (1976) Coconut hybridisation by the Polycaps and Mascopol systems. Principes 20:136-147
2. Manthriratna M (1965) Coconut pollen. Ceylon Coconut Quart 16:102-110
3. Marechal H (1928) Observation and preliminary experiments on the coconut with a view to developing improved seed-nuts for Fiji. Agric J Fiji 1:16-45
4. Nucé de Lamothe M de and Rognon F (1972) The production of hybrid seed by controlled natural pollination. Oléagineux 27:483-487

A brief history of coconut tissue culture

JENNET BLAKE
Unit for Advanced Propagation Systems, Department of Horticulture, Wye College, University of London, Wye, Ashford Kent TN25 5AH, UK

1. Introduction

The history of coconut tissue culture can be traced back to the 1960s when Dr. de Guzman started her work on embryo culture. Since then, there have been review papers which give more detail than is possible in this paper. They include general reviews emphasising vegetative propagation by Pannetier and Buffard-Morel [13] and Blake [5], as well as one on anther culture by Thanh-Tuyen [19]. There are two further reviews on somatic embryogenesis, by Verdeil and Buffard-Morel [20] and by Blake and Hornung [6], currently in press.

This review does not attempt complete coverage of the subject, but explores two areas in more depth to demonstrate how early work has been developed up to the present time, and to indicate how current activities may contribute to the production of improved breeding stock.

2. Embryo culture

The earliest attempts at coconut tissue culture were actually a form of 'embryo rescue' using the embryo of the Makapuno, a speciality coconut in the Philippines. Instead of the usual formation of endosperm and coconut water, the endosperm of the Makapuno coconut remains as a jelly, filling the entire cavity. The Makapuno coconuts are found as a small proportion of normal fruit on a few palms, generally of the 'Laguna Tall' ecotype. Such trees are highly prized since the Makapuno coconuts are an expensive delicacy. However, the occurrence of such coconuts appeared to be a chance event. Although the Makapuno coconut contains an apparently normal embryo, the embryo never germinates *in situ*.

C. Oropeza et al. (eds.), Lethal Yellowing: Research and Practical Aspects, 195–201.

In a series of papers (reviewed by Rosario and Guzman [18]), de Guzman and her co-workers described their success in germinating Makapuno embryos *in vitro*. The plants were established and yielded a higher proportion of Makapuno nuts than the parent. However, a proportion of normal coconuts were still produced. Hand pollination was carried out on the plants obtained from Makapuno embryos, and the resulting embryos were again cultured *in vitro*. Plants from these second generation embryos produced over 90% Makapuno-type coconuts.

The reason for the production of Makapuno nuts became clear at a later date. In normal nuts the enzyme α-D-galactosidase converts galactomannans to water-insoluble mannans which are deposited in the endosperm, leaving the centre of the nut filled with coconut water. This enzyme is deficient in the Makapuno nuts so that the endosperm remains as a jelly and fills the whole nut cavity [11, 12]. Such jelly coconuts are also found in other countries, and some commercial exploitation is now taking place. However, plantlets from Makapuno embryos can still command a high price [17].

There is considerable interest in embryo culture *per se*, since a successful technique would lead to an easier and safer method for the exchange of germplasm, as well as ultimately aiding conservation through cryopreservation of embryos. De Guzman's pioneering early work has been followed in a number of laboratories where zygotic embryos have been successfully cultured [1, 3, 10, 15]. The technique is not difficult to perform and up to 95% of cultured embryos can be established in soil (Rillo pers. comm.). Assy Bah *et al.* [3] found it necessary to remove the enlarged haustorium before field planting, but many laboratories find that the haustorium does not enlarge during *in vitro* culture.

Embryos can be transported successfully around the world using mail or rapid delivery systems [16]. In one method, they are excised at source, sterilised and placed in liquid in sealed vials for despatch. The liquid can be sterile water [2] or dilute sucrose and coconut water. As conditions during transport will be anaerobic, it is preferable to use a medium which does not encourage active growth of the embryo. An alternative successful method of transport is to remove a core of endosperm leaving the embryo *in situ.* With adequate sterilisation before and after excision from the core, the embryo germinates successfully. This method may be preferable since vials sometimes lose liquid during transport. Embryos which are not fully mature can be used and may germinate as well or better than mature embryos.

Successful medium-term cryopreservation of zygotic embryos has been reported by Assy-Bah and Engelmann [4] who obtained 100% recovery after storage for 5 months, although this was reduced to 51% after storage for 12 months.

Cultured embryos may be used for screening germplasm for resistance or tolerance to disease. Early work on this aspect was carried out by Rillo [14] with cadang-cadang disease, but has now ceased.

Micropropagation from zygotic embryos has been attempted in several laboratories (see review by Blake [5]). This would be a useful technique for multiplying elite hybrid embryos, particularly if direct secondary embryogenesis could be induced without the intervention of a callus phase. So far, this has not been reported. A successful technique of regeneration has now been developed from cultured plumules excised from mature zygotic embryos (Hornung, this volume).

3. Clonal propagation

Conventional breeding programs can only slowly improve the stock of coconut germplasm. Such programs have produced hybrids, usually dwarf x tall crosses, which have given considerable increases in yield. Many of these have come from the breeding program in the Ivory Coast which has produced the well-known hybrid, 'PB-121' or 'Mawa', which is a 'Malayan Dwarf' x 'West African Tall' cross. Although this hybrid has been widely planted, it is not successful in some locations and many countries wish to produce their own hybrids. Jamaica has planted the 'Maypan', a hybrid of the 'Malayan Dwarf' x 'Panama Tall'. This hybrid has been successful because it shows resistance to lethal yellowing which had devastated the native 'Jamaica Tall' population. A method of clonal propagation of parental breeding stocks would enable breeders not only to improve yields, but also to select for disease resistance or tolerance.

Many research projects have been directed towards developing a method of clonal propagation. Despite this, success has been very limited, but, where clonal plants have been produced, they appear to have come through a process of somatic embryogenesis. The first stage is the production of a callus, which is also termed 'calloid' as there is only partial dedifferentiation before regeneration. A true callus is sometimes formed but never regenerates. In attempts to propagate from mature trees, the original explants for the production of callus, or calloid, can be root, leaf, or inflorescence, but only leaf [8, 9] and inflorescence [7, 21] have given rise to regeneration.

Inflorescence explants are taken from young rachillae cut into slices, 1-2 mm thick, and placed onto a nutrient medium containing 2,4-D and activated charcoal. They are incubated in the dark at 29-30°C and dedifferentiation can be observed within 8-12 weeks, with the production of calloid from the flower meristem areas. This calloid consists of files of cells, apparently similar to the production of wound tissue in dicotyledons, though not normally seen in monocotyledons. Meristematic nodules are formed at the surface, and a considerable degree of multiplication of the nodules can occur. Although other auxins have been tested, 2,4-D in the presence of activated charcoal is most commonly used to produce calloid. Somatic embryos are formed when the concentration of 2,4-D is altered: an early increase is recommended by Verdeil *et al.* [21] with a later reduction. The changes suggested and their timing are often not specified, although somatic embryo development is associated with a decrease in auxin.

Two types of calloid, or callus, are produced, as shown by Verdeil *et al.* [21]. The most common is the multicellular calloid producing files of cells with regeneration at the surface. However, there is also a unicellular route (Buffard-Morel *et al.*, this volume). The time required from culturing the original explant to production of somatic embryos is not consistent between laboratories. At Wye College, somatic embryos have been obtained within 9 months, whereas at ORSTOM/CIRAD the same stage has been reached after about 2 years. This time difference may be due to the different types of calloid, or callus, being produced.

Even with an in-depth study of all the papers published in this area [5, 13, 20], it is not possible to draw conclusions as to the mechanism which is operating to produce calloid and somatic embryos. One problem is that no publication has given an unequivocal protocol, and where a protocol is suggested, the details are often incomplete. This seems to be because there is a lack of reproducibility and the authors cannot consistently repeat their own protocol, and therefore hesitate to publish the full details. For example, at Wye College, an initial protocol was developed which gave 50 or more plantlets from 20 mother palms over a period of about 3 years. Since that early success, the protocol has worked less well and improvement has not occurred, despite testing a wide range of growth regulators and environmental and nutrient conditions. Small numbers of plantlets are still regenerated from inflorescence tissue, but these are not as normal as those produced previously. ORSTOM/CIRAD also have recently obtained plantlets through somatic embryogenesis [21].

A major problem is that most development is abnormal, but this is not because the calloid fails to proceed through somatic embryogenesis. This is shown by the fact that somatic structures are produced which are equivalent to those developed on zygotic embryos. Shoot meristems occur only occasionally, but leaves and roots are frequently seen and are often fasciated. It could be suggested that these have been produced through organogenesis, but this seems unlikely since the most commonly produced structures resemble cotyledonary or haustorial tissue [6]. These structures may resemble the zygotic haustorium, or be embryo-shaped or just form a proliferative mass of white spongy tissue.

One of the important requirements is to control the balance of development between shoot, root and cotyledonary-type tissue. We know that high levels of auxin produce callus, or calloid, that moderate levels induce roots and that cotyledonary tissue occurs at low levels, or when auxin is omitted. As yet, we have no real indication of how to retain and develop the shoot meristem which is present in the somatic embryo at an early stage. Until there is a greater understanding, and hence a greater control, of embryogenesis, it will not be possible to define a reliable protocol for clonal propagation of coconut.

4. Conclusions

Many exciting possibilities lie ahead for coconut tissue culture, but it will take much time and persistence before these possibilities can be realised. A reliable method of clonal propagation is one of the urgent requirements so that breeders can multiply selected parental lines. Homogeneous breeding material produced through pollen culture may become possible in the longer term. Both these systems rely on obtaining a greater understanding of embryogenesis and this is now the aim of a collaboartion between the Albay Reseearch Centre (Philippines), CICY (México), CIRAD/ORSTOM (France), Hannover University (Germany), IDEFOR (Ivory Coast) and Wye College (UK), funded by the Commission of the European Communities. They are becoming linked to investigate fundamental aspects of embryogenesis which underpin the practical exploitation of this technique.

Conservation and exchange of coconut germplasm is an aspect which has received considerable attention, and is an area to which tissue culture can contribute through the adoption of cryopreservation techniques and subsequent growth and multiplication of tissues.

Although it is tempting to consider the possibilities of genetic modification, this must remain well in the future for coconut, as indicated at a recent conference in the Philippines. However, once the more basic work has become reliable, genetic modification will be able to contribute to altering, for example, the constituents of coconut oil, but, more importantly, should allow for the production of palms with resistance or tolerance to disease.

Improvement of the coconut will always be of importance to the countries where it is grown. Even if temperate countries are able to produce their own replacement for lauric acid, there will always be a high requirement for coconut products in the tropical areas of the world where the coconut is not only a subsistence crop for small farmers, but is also an important part of the diet for the general population.

Acknowledgements

I am grateful to the Deutsche Gesellschaft für Technische Zusammenarbeit (GTZ) for supporting coconut research at Wye College over many years. Currently, financial support also is being received from the European Community and the UK Overseas Development Administration. I extend my thanks to Mr. Hugh Harries who kindly amended the script and presented the paper for me.

References

1. Ashburner GR, Thompson WK, Maheswaran G and Burch JM (1991) The effect of solid and liquid phase in the basal medium of coconut (*Cocos nucifera* L.) embryo cultures. Oléagineux 46:149-152
2. Ashburner GR, Thompson WK, Richards D and Halloran GM (1993) Collecting and transporting coconut germplasm using embryo culture techniques. In: BC Imrie and JB Hacker, eds. Focussed Plant Improvement: Towards Responsible and Sustainable Agriculture. Proc 10th Aust Plant Breed Conf April 1993 Vol 2, pp 9-10. Canberra: Australian Convention and Travel Service
3. Assy Bah B, Durand-Gasselin T, Engelmann F and Pannetier C (1989) The *in vitro* culture of coconut (*Cocos nucifera* L.) zygotic embryos - revised and simplified method for obtaining coconut plantlets suitable for transfer to the field. Oléagineux 44:515-523
4. Assy-Bah B and Engelmann F (1993) Medium-term conservation of mature embryos of coconut. Plant Cell Tiss Organ Cult 33:19-24

5. Blake J (1990) Coconut (*Cocos nucifera* L.). In: YPS Bajaj ed. Biotechnology in Agriculture and Forestry, Vol 10 Legumes and oilseed crops I, pp 538-554. Berlin: Springer-Verlag
6. Blake J and Hornung R Somatic embryogenesis in coconut. In: SM Jain, ed. Somatic embryogenesis in woody plants. Dordrecht: Kluwer, in press
7. Branton RL and Blake J (1986) Clonal propagation of coconut. In: E Pushparajah CP Soon, eds. Cocoa and Coconuts: Progress and Outlook, pp 771-780. Kuala Lumpur: Incorporated Society of Planters
8. Buffard-Morel J, Verdeil JL and Pannetier C (1992) Embryogenèse somatique du cocotier (*Cocos nucifera* L.) à partir d'explants foliaires: étude histologique. Can J Bot 70:735-741
9. Karunaratne S, Gamage C and Kovoor A (1991) Leaf maturity, a critical factor in embryogenesis. J Plant Physiol 139:27-31
10. Karunaratne S, Kurukulaarachchi C and Gamage C (1985) A report on the culture of embryos of dwarf coconut. *Cocos nucifera* L. var. *nana in vitro*. Cocos 3:1-8
11. Mujer CV, Ramirez DA and Mendoza EMT (1984) α-D-galactosidase deficiency in coconut endosperm: its possible pleiotropic effects in makapuno. Phytochem 23:893-894
12. Mujer CV, Ramirez DA and Mendoza EMT (1984) Coconut α-D-galactosidase isoenzymes: isolation, purification and characterization. Phytochem 23:1251-1254
13. Pannetier C and Buffard-Morel J (1986) Coconut palm (*Cocos nucifera* L.). In: YPS Bajaj, ed. Biotechnology in Agriculture and Forestry Vol 1 Trees, pp 430-450. Berlin: Springer-Verlag
14. Rillo EP (1985) Cadang-Cadang Research. *In vitro* mechanical inoculation of different coconut cultivars and hybrids with ccRNA. Annu Rep Philipp Coconut Authority 1985, p 93
15. Rillo EP and Paloma MBF (1990) Comparison of 3 media formulations for *in vitro* culture of coconut embryos. Oleagineux 45:319-323
16. Rillo EP and Paloma MBF (1991) Storage and transport of zygotic embryos of *Cocos nucifera* L. for *in vitro* culture. Plant Genet Resour Newsl 86:1-4
17. Rillo EP and Paloma MBF (1992) *In vitro* cultures of Makapuno coconut embryos. Coconuts Today 9:90-108
18. Rosario AG del and Guzman EV de (1982) The status of plant tissue culture in the Philippines. In: AN Rao, ed. Tissue Culture of Economically Important Plants, pp 293-294. Singapore: COSTED and ANBS
19. Thanh-Tuyen NT (1990) Coconut (*Cocos nucifera* L.): anther culture. In: YPS Bajaj, ed. Biotechnology in Agriculture and Forestry Vol 10: Legumes and Oilseed Crops I, pp 555-568. Berlin: Springer-Verlag
20. Verdeil JL and Buffard-Morel J Somatic embryogenesis in coconut (*Cocos nucifera* L.). In: YPS Bajaj ed. Biotechnology in Agriculture and Forestry, Somatic Embryogenesis Vol 2. Berlin: Springer-Verlag, in press
21. Verdeil JL, Huet C, Grosdemanges F, Rival A and Buffard-Morel J (1992) Coconut (*Cocos nucifera* L.) somatic embryogenesis: obtention of several clone ramets. Oléagineux 47:465-469

Initiation of callogenesis in coconut palm (*Cocos nucifera* L.)

ROLAND HORNUNG
Unit for Advanced Propagation Systems, Department of Horticulture, Wye College, University of London, Wye, Ashford Kent TN25 5AH, UK

1. Introduction

Although *in vitro* propagation of coconut palms has been researched for more than two decades [1, 7], this species remains recalcitrant and is still solely propagated sexually. Success in producing a very limited number of plantlets through *in vitro* propagation by somatic embryogenesis has been reported [2], but it is disappointing that during these two decades, only two or three vegetatively propagated plantlets have been established *ex vitro* in the field. It is not stated whether the plantlets produced were truly clonal or not. This is a small return for the level of input into coconut tissue culture research. Apparently, there is still no protocol available for the reproducible *in vitro* production of clonal plants on a large scale.

Recent studies with immature inflorescences, immature and mature zygotic embryos and plumules excised from mature zygotic embryos are reported in this paper. Emphasis is placed on the production of calloid from these various tissues when exposed to different auxins. The term calloid is defined as partially dedifferentiated callus, and, as it is the first stage in the process of somatic embryogenesis, its production is of major significance in the culture of coconut.

2. Materials and methods

Three different sources of explant tissues were used. Immature inflorescences and zygotic embryos obtained from Jamaica, the Philippines and México, were transported by air to the UK. Plumular tissue derived from the mature zygotic embryos constituted the third explant type investigated.

C. Oropeza et al. (eds.), Lethal Yellowing: Research and Practical Aspects, 203–215.

2.1 *Immature inflorescence*

Immature inflorescences from palms of *Cocos nucifera* L., 'Malayan Dwarf' and 'Laguna Tall', were received from the countries cited above. In each case, the developmental stage of an immature inflorescence was determined by its position within the sequence on the palm. Thus, the inflorescence referred to as +1 represents the most recently opened, while -1 is the next inflorescence to open. In the case of -4, for example, the inflorescence would open *in situ* approximately 4 months later. After removal of the outer and inner spathes, the rachillae were aseptically cut into 0.5-2.0 mm long segments, depending on the age of the inflorescence, and cultured on solid media.

2.2 *Mature zygotic embryos and plumular tissue*

Cores of endosperm containing either immature (7-9 months after pollination) or mature zygotic embryos (11-12 months after pollination) were removed from the coconuts in the country of origin. In the UK, the cores were sterilised in 37.5% commercial bleach (Brobat, Jeyes UK, 4.6% NaOCl) for 10 min followed by three rinses with sterile water. The embryos were cut from the endosperm and sterilised for a second time in 10% commercial bleach for 5 min, followed by three rinses with sterile water. When plumular tissue was required, this was excised from the sterile mature embryos using a dissecting microscope after they had been cultured in liquid Murashige and Skoog medium (MS) [5] supplemented with 4% sucrose. Embryos and plumules were explanted onto solid media containing MS macro- and micro-nutrients prior to receiving growth regulator treatments.

2.3 *Media composition*

The calloid induction media were either a revised Eeuwens [3] (Y3R), Murashige and Skoog [5] (MS, Imperial Laboratories), Boston Fern Multiplication (BF, Flow Laboratories) or Woody Plant Medium (WP, Imperial Laboratories). These were supplemented with 4% sucrose, 0.25% activated charcoal, gelrite (Schweizerhall, UK) and plant growth regulators (2-iP, BAP, NAA, dicamba, fluroxypyr acid, 2,4,5-T, picloram and 2,4-D). All media were adjusted to pH 5.8 prior to the addition of gelling agent and activated charcoal, and were autoclaved at 121°C (1.6 kg cm^{-3}) for 20 min. Unless otherwise stated, the cultures were maintained at

29±1°C in darkness. Explants were transferred to fresh medium at intervals of 4 weeks.

Much of the data obtained, particularly that with immature inflorescences, was highly variable within treatments, and no statistical analyses were performed.

3. Experimental and results

3.1 Immature inflorescence tissue

Four experiments were performed to investigate the effects of plant growth regulator treatments on the production of calloid using explants from immature inflorescences at different developmental stages. The influence of genotype and its interactions with the other parameters was also investigated.

3.1.1 Effect of auxin concentration in the induction medium

The effect of five concentrations of 2,4-D from 10^{-4} to $4x10^{-4}$ M on the production of calloid from inflorescences of 'Laguna Tall' and 'Malayan Dwarf' at the same developmental stage (-5) was investigated. Forty replicate tubes were used for each treatment, with each tube containing four explants. The percentage of explants which had produced calloid after 6 months culture on medium Y3R was assessed.

Table 1. The effect of 2,4-D concentration (10^{-4}M to $4x10^{-4}$M) on the production of calloid by immature inflorescence explants from two genotypes ('Laguna Tall', 'Malayan Dwarf'; stage -5). Values are percentages of explants bearing calloid.

2,4-D conc (M)	Laguna Tall: palm number					Malayan Dwarf: palm number				
	1	2	3	4	5	1	2	3	4	5
10^{-4}	1.25	0	0	0	0	0	0	0.62	0	0
$1.5x10^{-4}$	2.5	0t	1.87	0	0	0t	0	0	0t	0
$2x10^{-4}$	0T	0t	0.62	0	0	0t	0	0	0	0t
$3x10^{-4}$	0T	0T	0T	2.5	0	0T	0T	0T	0T	0
$4x10^{-4}$	0T	0T	0	0T	0	0T	0T	0T	0T	0T

0 = no calloid; t = toxicity and T = severe toxicity.

Table 1 shows that the induction of calloid on the two genotypes was variable and no consistent pattern was found relating calloid formation with different 2,4-D concentrations within the range used. Auxin concentrations higher than $3x10^{-4}$ M caused severe toxicity with nearly all inflorescences used in this and other experiments. There was considerable exudation from the tissues when concentrations above $2x10^{-4}$ M 2,4-D were applied.

3.1.2 Effect of developmental stage on the response of immature inflorescence tissue at varying 2,4-D concentrations

Earlier studies had indicated that the auxin sensitivity and requirement of explants may be related to the developmental stage of the inflorescence of origin. This relationship was investigated in an experiment in which eleven inflorescences (-4 to -14) were taken from a single Malayan Dwarf palm. The five concentrations of 2,4-D and other experimental details were as in 3.1.1.

The results shown in Table 2 indicate that 2,4-D concentrations in excess of $2x10^{-4}$ M approached toxic levels for this genotype. Calloid was produced only with inflorescence of -8 and younger, and at the lower concentrations of 2,4-D. Successful cultures occurred randomly throughout the table and were too few to be conclusive (5 explants with calloid from over 5 000 explants which had not received toxic treatments).

Similar inconsistent results were obtained with two further genotypes but these are not reported here.

Table 2. Calloid production from explants taken from inflorescences at different developmental stages (-4 to -14) when exposed to varied concentrations of 2,4-D ($1\text{-}4x10^{-4}$ M). Values are percentages of explants with calloid.

2,4-D conc (10^{-4} M)	Developmental stage of the inflorescence										
	-4	-5	-6	-7	-8	-9	-10	-11	-12	-13	-14
1	0	0	0	0	1.25	0	0	0	0	0	1.25
1.5	0	0	0	0	0	0	0	0.62	0	0	0
2	0	0t	0	0	0	0	0	0	0	0	0
3	0T	0T	0T	0t	0T	0	0t	0t	0T	0t	0t
4	0T	0T	0T	0T	0T	0T	0T	0T	0T	0T	0T

0 = no calloid; t = toxicity and T = severe toxicity.

3.1.3 Combined effects of medium composition and 2,4-D concentration

During the early work at Wye College with immature inflorescences, a modified medium was developed (Y3R). Procedures have since changed and a re-examination of the suitability of this medium became necessary.

In addition to Y3R, Murashige and Skoog (MS), half strength MS (½ MS), Boston Fern Multiplication (BF) and Woody Plant (WP) media were used with 2,4-D at 10^{-4} and $2x10^{-4}$ M. The tissue was taken from two different palms (Malayan Dwarf 1 and 2) at two developmental stages of the inflorescences (-6, -9). Other procedures were as described above. Values for percentages of cultures which produced calloid are presented in Table 3. The data do not reveal any special nutrient requirement for inducing calloid. The highest percentage of calloid was obtained with Palm 2 when the explants were exposed to BF medium.

Table 3. Comparison of calloid production on explants of immature inflorescence from two 'Malayan Dwarf' genotypes, when cultured on different media.

medium	inflorescence stage: replicate 1				inflorescence stage: replicate 2			
	-6		-9		-6		-9	
	2,4-D concentration ($x10^{-4}$ M)							
	1	2	1	2	1	2	1	2
Y3R	0	0	0.62	1.25	0	0	0	0
MS	0	0	0	1.87	0	0	0	0.62
½MS	0	0T	0	0	0	0	0	0
BF	0	0	1.25	0	1.25	0	1.25	1.87
WP	0	0t	0	0	0	0	0	0

0 = no calloid; t = toxicity and T = severe toxicity.
Y3R = Revised Eeuwens; MS = Murashige and Skoog; BF = Boston Fern Multiplication; WP = Woody Plant.

No calloid was obtained with tissue from Palm 2, although Palm 1 produced low amounts of calloid at both auxin levels with the younger tissue. Data in Table 3 suggest that half strength MS and WP media were not suitable.

The -6 inflorescence taken from Palm 2 only produced calloid on BF medium containing 10^{-4} M 2,4-D, whereas twice the amount of auxin also gave a positive result with the -9 tissue of the same palm. Explants of the younger inflorescences from Palm 1, which were cultured on BF medium containing 10^{-4} M 2,4-D, produced calloid, whilst BF supplemented with twice the amount of auxin resulted in no calloid induction. An analogous response was observed with the same explant source on MS medium.

3.1.4 Effect of other synthetic auxins as alternatives to 2,4-D for the induction of calloid

An inflorescence (-6) from a 'Malayan Dwarf' palm was used to provide explants to compare the effectiveness of five plant growth regulators for inducing calloid. The regulators listed in Table 4 were each incorporated into the media (MS) at three concentrations (10^{-4} M, $2x10^{-4}$ M and $3x10^{-4}$ M). All other experimental procedures were as previously described and the percentages of explants which had produced calloid after 6 months are recorded in Table 4.

Table 4. Comparison of the induction of calloid on inflorescence explants of 'Malayan Dwarf' cultured on media containing various synthetic auxins each at three concentrations.

Growth regulator	Growth regulator concentration (10^{-4} M)					
	1	2	3	10	20	30
2,4-D	0.62	0	0T	-	-	-
2,4,5-T	0.62	0t	0T	-	-	-
fluroxypyr	0T	0T	0T	-	-	-
dicamba	0T	0T	0T	-	-	-
picloram	0	0T	0T	-	-	-
NAA	-	-	-	0	0t	0t

- = not tested, 0 = no calloid, t = toxicity and T = severe toxicity.

In this experiment, positive results were only obtained at the lowest concentration of 2,4-D and 2,4,5-T. Even here, the values were low (0.62%). Although these results suggest that fluroxypyr acid, dicamba and picloram were used at too high a concentration, further experiments with lower concentrations were still not successful.

3.2 Immature zygotic embryos

Two experiments were performed to determine the effects of plant growth regulator treatments on the production of calloid from immature zygotic embryos.

3.2.1 *Effect of auxin in induction media*

The effect of 5 different concentrations of 2,4-D from 10^{-4} M to $5x10^{-4}$ M on the production of calloid from immature zygotic embryos from 'Malayan Dwarf' was investigated. The procedures were as described in the materials and methods and 20 replicate tubes were used for each treatment, each tube containing one embryo. The percentages of explants which had produced calloid after 4 months of culture on MS medium were assessed.

The results shown in Table 5 indicate that the concentration of 10^{-4} M 2,4-D allowed considerable germination but was not strong enough to induce calloid, whereas the next two higher auxin concentrations induced some dedifferentiated tissue as well as allowing germination. Levels of $4x10^{-4}$ M 2,4-D produced the highest amount of calloid. Further increase of the 2,4-D concentration led to a decrease in calloid formation.

Table 5. The effect of 2,4-D concentration on the percentage production of calloid from immature zygotic embryos ('Malayan Dwarf').

Explant response (%)	2,4-D concentration ($x10^{-4}$ M)				
	1	2	3	4	5
calloid	0	0	0	10	5
germination	75	80	60	35	25
calloid + germination	0	5	5	5	5
expansion	20	10	20	15	20
no response	5	5	15	35	45
total calloid	0	5	5	15	10

Exposure of the embryos to increasing 2,4-D concentrations led to a decrease in germination. All 2,4-D concentrations higher than 10^{-4} M produced calloid in inverse proportion to the proportion of embryos germinating. Only the two highest 2,4-D concentrations induced calloid on embryos which did not germinate at the same time. The proportion of embryos showing no response increased with increasing 2,4-D concentration.

3.2.2 *Effects of different plant growth regulators in induction media*

Further information was obtained about the response of zygotic embryos exposed to plant growth regulators in an experiment which compared the effects of 2,4-D, picloram, 2,4,5-T and fluroxypyr acid. Using the procedures of 3.1.1, these regulators were each tested at 3 concentrations, $2x10^{-4}$, $4x10^{-4}$ and $6x10^{-4}$ M. The development of calloid was assessed over a period of 24 weeks. The assessments were made after 8, 10 and 12 weeks, and subsequently at 4 week intervals.

The first calloid was produced after 10 weeks incubation, to the greatest extent with 2,4-D (Table 6), but also with the highest concentration of picloram. After 12 weeks, strong evidence of calloid formation was seen with all treatments except fluroxypyr acid, where no response occurred until 16 weeks of culture. The fluroxypyr acid result was atypical in that only the lowest concentration was effective and that comparable levels of other growth regulators were too low to give a response.

Table 6. Calloid production (%) from immature zygotic embryos when exposed to different plant growth regulators (2,4-D, 2,4,5-T, fluroxypyr acid and picloram) at varied concentrations ($2\text{-}6x10^{-4}$ M), for 24 weeks after culturing.

Week	Growth regulator concentration ($x10^{-4}$ M)											
	2,4-D			2,4,5-T			fluroxypyr acid			picloram		
	2	4	6	2	4	6	2	4	6	2	4	6
8	0	0	0	0	0	0	0	0	0	0	0	0
10	0	5	10	0	0	0	0	0	0	0	0	5
12	0	15	15	0	10	5	0	0	0	0	5	15
16	0	15	20	0	10	5	5	0	0	0	10	15
20	0	15	20	0	10	5	5	0	0	0	10	20
24	0	15	20	0	10	5	5	0	0	0	10	20

The results of visual assessment showed consistently increased calloid production with increasing concentrations of 2,4-D and picloram. 2,4-D gave the highest values, but these were only marginally higher than picloram. Fluroxypyr acid induced its response at a lower concentration than the other test compounds and 2,4,5-T may also have been supraoptimal at $6x10^{-4}$ M.

3.3 Plumular tissue

In many investigations with coconut zygotic embryos, the greatest responses have been obtained with very young tissue. It therefore appeared logical to experimentally use the plumules from mature zygotic embryos.

3.3.1 Production of calloid from plumular tissue

In preliminary experiments, it was observed that plumular tissue could tolerate 2,4-D concentrations which were some five times greater than those tolerated by immature inflorescence explants (Hornung unpubl.). The range chosen was therefore $2x10^{-4}$ to $5x10^{-4}$ M.

For each treatment, 40 replicate culture tubes were used, each containing plumular tissue excised from one mature zygotic embryo as described in 3.2.

The data in Table 7 show a rate of calloid production of more than three times that for the zygotic embryo explants and infinitely greater than that for the responsc with immature inflorescences normally achieved in these laboratories. The results are also notable in respect to the sensitivity of the system to 2,4-D levels. As the 2,4-D concentration was increased, the germination of the plumule was suppressed in inverse proportion to the rate of calloid production, with over 70% of cultures producing calloid at the two highest concentrations. Although the results represent only one experiment, some 1 000 plumules have now been cultured and the findings repeated.

Table 7. Percentage calloid production from plumules when exposed to varied concentrations of 2,4-D ($2x10^{-4}$ M to $5x10^{-4}$ M) after 8 weeks of culture.

Explant response (%)	2,4-D concentration ($x10^{-4}$ M)			
	2	3	4	5
calloid	0	15	72.5	72.5
germination	65	45	5	2.5
calloid + germination	0	0	5	0
growth	35	40	17.5	25
total calloid	0	15	77.5	72.5

4. Discussion

A variety of factors have been shown to affect the calloid induction process from inflorescence tissue. The results indicated that the genotype of the palm which served as explant source is very important (Table 1). The results given here lack reproducibility within one ecotype and it has not yet been possible to obtain further explant material from the same palm at a later date. The environment is also an important factor, including growth conditions of the palm in the field, the conditions both in transit and during the *in vitro* procedure. Research performed in European laboratories mainly deals with the environment during the *in vitro* procedures. Different genotypes responded very inconsistently to the media containing various macro- and micro-nutrient compositions (Table 3) and also to the use of different plant growth regulators (Table 4). This might be attributable to the genotype, the environmental conditions in the field (season *etc.*), or during the *in vitro* procedure as well as to the developmental stage of the explant (Table 2), which produced inconsistent data although derived from a single genotype.

Although only small amounts of calloid have been produced from many explants, the uniformity of this visually assessed calloid was very low. On some occasions, a batch of immature inflorescences has produced a good yield of calloid but did not occur in any of the present series of experiments. The lack of uniform response is a severe handicap to the use of inflorescence tissue.

If immature inflorescence tissue is to be satisfactory for calloid induction, then:

a) palms being utilised as explant sources should be grown under more controlled conditions particularly in relation to water supply, and

b) individual palms must be screened for their ability to produce calloid, with immature inflorescence harvested non-destructively [6], so that successful genotypes can be used for further studies.

Production of calloid from immature zygotic embryos in high and consistent proportions was reported by Karunaratne and Periyapperuma [4]. The results obtained here gave consistent calloid production but only at a low level.

If zygotic embryos are to be satisfactory for calloid induction, then:

a) information about the parents of the embryo must be available, and

b) markers must be found for parental characteristics which can be used to predict the performance of the clonal plants obtained from the zygotic embryos.

Culture of plumular material has considerable potential in offering a method for the large-scale production of calloid for research on embryogenesis in coconut.

Like the procedures involving the use of entire zygotic embryos (3.2), the plumules can only produce clones of palms with unknown performance, but if zygotic embryos are produced from elite parents, then clones from these embryos may show many of the characteristics of the parents.

Inflorescence explants have, on rare occasions, produced up to 30% of cultures with calloid. This level of induction was not achieved in the course of the experiments reported here. When calloid does occur, it appeared to be a random occurrence. This contrasts with other tissues and particularly with explants of plumular tissue excised from mature embryos. Results obtained with immature zygotic embryos were much less encouraging in respect to the level of calloid production, but were satisfactory in term of reproducibility and did not vary from consignment to consignment.

A further difference in the behaviour of these explant materials is the time taken in culture to produce calloid. Plumular explants require as little as 4 weeks before calloid appears, compared with 10 weeks with immature zygotic embryos and 12 weeks [1] or 8 months [8] with immature inflorescence. This reduction in time is seen as an advantage because:

a) the tissue is exposed to high levels of plant growth regulators for shorter periods,

b) when calloid is produced over a shorter period, the decision about the timing of future treatments is easier,

c) costs are reduced with fewer transfers and the shorter period of use of facilities, and

d) there is less chance of infection with fewer transfers over a shorter time.

The assessment of calloid from each of the three types of explant was made visually and it is not yet possible to relate the ability of the calloid to the production of normal somatic embryos. Such information must await the availability of "markers" for early embryogenesis, hopefully to be undertaken soon in the laboratories of Prof. H.J. Jacobsen in Hannover, Germany.

Sound evidence concerning the relative responses of the various tissues to different plant growth regulators would be helpful in optimising calloid production. Unfortunately, the results obtained with inflorescence are not consistent enough to make this possible where the substituted phenoxyacids 2,4-D and 2,4,5-T are both partially effective. With zygotic embryos, more information is available and the picolinic acid, picloram, was almost as effective as 2,4-D. Investigation of plumular tissues carried out with picloram, as yet incomplete, are also encouraging.

5. Conclusions

The ultimate success or failure of the micropropagation of elite coconut palms will depend upon the development of a regeneration system which avoids the production of abnormal plantlets. Factors such as the plant growth regulators used, their concentration and period of treatment and the type of explant will be very significant in this respect.

Acknowledgements

I am most grateful to Drs J. Blake and H.F. Taylor for their valuable guidance and encouragement throughout the course of this research. I appreciate the assistance of members of the Horticulture Department, Wye College. Plant material was kindly supplied by CICY, México, Philippine Coconut Authority Albay Research Center, Philippines and Coconut Industry Board, Jamaica. This research was supported by the EC and GTZ, and the British Council provided travel assistance.

References

1. Blake J (1990) Coconut (*Cocos nucifera* L.): Micropropagation. In: YPS Bajaj, ed. Biotechnology in Agriculture and Forestry Vol 10 Legumes and Oilseed Crops I, pp 538-554. Berlin: Springer-Verlag
2. Blake J and Hornung R Somatic Embryogenesis in Coconut. In: Jain SM, ed. Somatic Embryogenesis in Woody Plants. Dordrecht: Kluwer (in press)
3. Branton RL and Blake J (1983) Development of organised structures in callus derived from explants of *Cocos nucifera* L.. Ann Bot 52:673-678
4. Karunaratne S and Periyapperuma K (1989) Culture of immature embryos of coconut (*Cocos nucifera* L.) callus proliferation and somatic embryogenesis. Plant Sci 62:247-253
5. Murashige T and Skoog F (1962) A revised medium for rapid growth and bioassays with tobacco tissue cultures. Physiol Plant 15:473-497
6. Rillo EP (1989) A nondestructive technique for collecting immature coconut inflorescence for tissue culture. Phillip J Coconut Stud 14:16-17

7. Verdeil JL and Buffard-Morel J Somatic embryogenesis in coconut (*Cocos nucifera* L.). In: YPS Bajaj, ed. Biotechnology in Agriculture and Forestry. Somatic embryogenesis. Berlin: Springer-Verlag (in press)
8. Verdeil JL, Huet C, Grosdemange F and Buffard-Morel J (1994) Plant regencration from cultured immature inflorescences of coconut (*Cocos nucifera L.*): evidence for somatic embryogenesis. Plant Cell Rep 13 218-21

Initiation of somatic embryogenesis in coconut (*Cocos nucifera* L.)

J. BUFFARD-MOREL, J.L. VERDEIL, S. DUSSERT, C. MAGNAVAL, C. HUET and F. GROSDEMANGE.

Laboratorie des Ressources Genetiques et Amelioration des Plantes Tropicales, ORSTOM/CIRAD-CP, BP 5045, 34032 Montpellier Cedex, France

1. Introduction

The vegetative multiplication of high-performance individuals is a promising possibility for producing homogeneous planting material and improving the productivity and homogeneity of coconut plantations. Cloning would also result in more rapid exploitation of breeding results with rapid distribution of clones in breeding programs. Mastery of coconut regeneration by somatic embryogenesis would open the way to genetic transformation by introduction of genes bestowing resistance to certain diseases, thereby giving resistant plants.

In vitro culture remains the only approach to vegetative propagation of coconut and somatic embryogenesis appears to be the most promising technique.

The first results obtained by the principal teams working on the subject were promising (1980-1984). However, the scarcity of articles published on the subject between 1985 and 1988 illustrates the difficulties encountered by all teams in obtaining whole somatic embryos with shoot development. Between 1984 and 1988, only four teams achieved whole plant regeneration from somatic tissues [1, 2, 3, 10], but the ramets were apparently obtained from a limited number of cultures. The ramets obtained proved that coconut regeneration by somatic embryogenesis is possible, but simply has not been mastered. It is clear that coconut remains an extremely recalcitrant plant to *in vitro* culture.

The difficulties in coconut regeneration are numerous: intense browning of tissues linked to their high sensitivity to synthetic auxins (*e.g.* 2,4-D), considerable

C. Oropeza et al. (eds.), Lethal Yellowing: Research and Practical Aspects, 217–223.

heterogeneity in tissue reaction, low capacity for embryogenesis and caulogenesis and an extremely slow response *in vitro*.

Activated charcoal is often used in plant tissue culture for its promotion of somatic embryogenesis and for its potent adsorption of growth inhibitors secreted by the tissues, such as phenolic compounds. Given the high sensitivity of coconut tissues to synthetic auxins, the use of activated charcoal is essential for the maintenance of the cultures. Coconut calli have low embryogenic potential, and embryogenesis is often incompletely expressed, leading to fused or incomplete embryos without shoots. Coconut is also characterised by the slowness with which the different events occur and by the highly determined tissue component. These difficulties complicate the search for mediators of induction and expression of somatic embryogenesis.

By adopting an analytical approach (histological study, investigation of culture medium composition), the joint ORSTOM/CIRAD team at Montpellier acquired a clearer understanding of the reasons that embryogenesis was blocked or deviated. In this way, the Montpellier team obtained reproducible whole embryos. In total, approximately forty plantlets have been obtained from several clones belonging to different genotypes [11]. The different steps for development of somatic embryogenesis are described below.

2. Callus induction

The explants were received from the Ivory Coast (Marc Delorme Station, IDEFOR). This material was taken from adult individuals selected for their agronomic value but were predominantly from 'Malayan Yellow Dwarf' and the 'Malayan Yellow Dwarf' x 'West African Tall' ('PB-121') and 'Cameroon Red Dwarf' x 'West African Tall' ('PB111') hybrids created by CIRAD-CP. Two types of explants were used in the laboratory: young non-chlorophyllous leaves and immature inflorescences.

The basal medium consisted of Eeuwens inorganic nutrients [6] and the vitamins of Morel and Wetmore [9], supplemented with 30 gl^{-1} sucrose, 2,4-D (40, 50, 60, 70 mgl^{-1}) and 2 gl^{-1} of neutralised activated charcoal (Sigma). The medium was gelled with 7.5 gl^{-1} agar (Sigma). The pH was adjusted to 4.5 before autoclaving for 20 minutes at 120°C.

Calli appeared after four months of culture, with an optimum of 6-8 months. Histological studies identified sites of callus growth on both explant sources and provided a better understanding of the mechanisms involved in its growth. Neoformed tissues have two possible origins; a deep origin for calli obtained from leaf explants where the calli formed on undifferentiated perivascular sites, and a superficial origin, observed for calli forming on floral areas. Whatever the tissue origin, the growth and multiplication of calli were ensured by a peripheral meristematic zone frequently exhibiting cambium-like organisation. Callus multiplication could be maintained for long periods without any obvious histological change.

3. Initiation of somatic embryogenesis

In order to circumvent the difficulties encountered in coconut regeneration, homogeneous calli strains were obtained by the multiplication of the primary calli. In the presence of optimum 2,4-D concentration, callus strains were selected on the basis of their embryogenic competence. This material was suitable for improving the understanding of the induction of embryogenesis and for studying different factors that may promote embryogenesis.

3.1 *Histological study*

A histological study of embryogenic calli revealed two embryogenic pathways [4]:

a) a multicellular pathway which led to the formation of meristematic and epidermised structures obtained with low 2,4-D concentrations (40 to 60 mgl^{-1}, and 2 gl^{-1} charcoal). The first stage of development of these structures was characterised by the fragmentation of the cambium-like zone and the formation of complex meristematic structures followed by their epidermisation. Too low a concentration of auxin sometimes resulted in the incomplete or deviated embryonic structures (*e.g.* a haustorium only; with or without a root pole; foliaceous embryos).

b) a unicellular pathway which led to the appearance and individualisation of embryogenic cells isolated by a thick wall, with dense cytoplasm, a high nucleo-cytoplasmic ratio, a single large nucleolus, and starch and protein reserves. This second pathway was obtained in the presence of high 2,4-D concentrations (80-120 mgl^{-1} with 2-3 gl^{-1} charcoal) and led to the formation of typical pro-embryos characteristic of the initial stages of zygotic embryogenesis described by Haccius and Philip [8].

3.2 *Mineral nutrient requirements*

Specific nutritional requirements during initiation of embryogenesis were observed irrespective of the embryogenic process. This study revealed greater NH_4^+, Ca^{2+}, Mg^{2+} and sucrose absorption per gram of dry weight during embryogenesis. These specific nutritional requirements were accompanied by an increase in cellular protein reserves.

3.3 *Hormonal study*

Charcoal strongly adsorbs the growth inhibitors secreted by *in vitro* tissues, such as polyphenolic compounds. Fridborg and Eriksson [7] suggested that charcoal might also adsorb certain growth regulators, particularly synthetic auxins. The use of activated charcoal introduces a further source of variability in culturing, and under certain conditions of medium preparation, may be responsible for variable tissue performance in callogenesis and non-reproducible results.

A quantitative study of 2,4-D adsorption was carried out in the Montpellier laboratory using HPLC in a solid medium with 2,4-D detection under UV light. Under our culturing conditions, the results confirmed high 2,4-D adsorption by the activated charcoal. Thus, with 4.5 x 10^{-4} M 2,4-D and 3 gl^{-1} activated charcoal, 0.47% of the total auxin initially present in the medium remained available 21 days after the media were prepared.

A kinetic study showed that 2,4-D adsorption by the activated charcoal was a gradual process. The time taken for the 2,4-D level to stabilise depended on the initial 2,4-D concentration and the activated charcoal content. Using 3 gl^{-1} activated charcoal and 100 mgl^{-1} 2,4-D, the free 2,4-D/adsorbed 2,4-D equilibrium was reached 21 days after media preparation.

4. Somatic embryo maturation

According to most authors, the essential factor in embryogenic expression is the gradual drop in 2,4-D concentration in relation to the auxin level used for induction of embryogenesis. However, the initial concentration from which this gradual reduction occured was the essential factor; it ensured good initiation of embryogenesis likely to yield whole embryos. If the drop in auxin rate was too rapid, the maturation of the embryogenic structures usually led to incomplete or deviated forms.

The frequent deviation of embryonic morphogenesis could be linked to a hormonal imbalance [5]. Addition of an exogenous cytokinin (BAP) resulted in complete embryogenesis. A strictly defined sequence of events had to be followed to obtain complete embryogenesis: haustorium elongation, cauline pole differentiation and root pole differentiation. Without exogenous cytokinins, premature differentiation of the root pole led to a morphological dead-end, with irreversible inhibition of cauline pole differentiation.

5. Embryoid conversion

Conversion of embryos with shoot emission was carried out on an auxin-free medium. Establishment of the cauline meristem preceded root-pole differentiation. The root could develop spontaneously, indicating the presence of a functional bipolar axis, but in most cases addition of NAA was necessary to obtain roots. The haustorial tissue located opposite the stem meristem contained starch reserves. The location of starch grains, which is a clear indicator of embryo polarity, was identical to that found in a zygotic embryo.

6. Conclusions

Clones of several genotypes have been obtained in our laboratory. However, the number of ramets regenerated per clone remains low due to the lack of an intense multiplication phase. The main objective of future work will be to determine the conditions required for achieving adventive embryogenesis. In 1992, embryogenic suspensions were established with competent friable calli. This technique has the potential for the mass production of individual embryos in synchronous growth.

Fig. 1. a: Embryogenic structure (x6). b: Initiation of adventive embryogenesis from embryos with developed haustoria (x3). c: Shootlets from a clump of somatic embryos (x3). d: Somatic embryo conversion with first leaves and root emergence on an auxin-free medium (x3). e: Vitroplant from somatic embryo (x1).

References

1. Bhalla-Sarin N, Bagga S, Sopory SK and Guha-Mukherjee S (1986) Induction and differentiation of callus from embryo of *Cocos nucifera* L. by IAA-conjugates. Plant Cell Rep 5:322-324
2. Branton RL and Blake J (1984) Clonal propagation of coconut palm. In: E Pushparajah and CP Soon, eds. Cocoa and Coconuts: Progress and Outlook, pp 771-780. Kuala Lumpur: Incorporated Society of Planters
3. Buffard-Morel J, Verdeil JL and Pannetier C (1988) Vegetative propagation of coconut palm through somatic embryogenesis, obtention of plantlet from leaf explant. In: G Durand, L Bobichon and J Florent, eds. Proceedings of the 8th International Biotechnology Symposium, Paris, p 117. Paris: Société Francaise de Microbiologie
4. Buffard-Morel J, Verdeil JL and Pannetier C (1992) Embryogenèse somatique du cocotier (*Cocos nucifera)* à partir de tissus foliaires: étude histologique. Can J Bot 70:735-741
5. Dublin P, Enjalric F, Lardet L, Carron MP, Trolinder N, and Pannetier C (1991) Estate crops. In: PC Debergh and RH Zimmerman, eds. Micropropagation Technology and Application, pp 337-361. Dordrecht: Kluwer
6. Eeuwens CJ (1976) Mineral requirements for growth and callus initiation of tissue explants excised from mature coconut (*Cocos nucifera*) and date (*Phoenix dactylifera*) palms cultured *in vitro.* Physiol Plant 36:23-28
7. Fridborg G and Erikson T (1975) Effects of activated charcoal on growth and morphogenesis in cell cultures. Physiol Plant 34:306-308
8. Haccius B and Philip VJ (1979) Embryo development in *Cocos nucifera* L.: a critical contribution to a general understanding of palm embryogenesis. Plant Syst Evol 132:91-106
9. Morel G and Wetmore RM (1951) Fern callus tissue culture. Am J Bot 38:141-143
10. Raju CR, Prakash Kumar P, Chandramohan M and Iyer RD (1984) Coconut plantlets from leaf tissue cultures. J Plantation Crops 12:75-81
11. Verdeil JL, Huet C, Grosdemange F and Buffard-Morel J (1994) Plant regeneration from cultured immature inflorescences of coconut *(Cocos nucifera* L.*)*: evidence for somatic embryogenesis. Plant Cell Rep 13:218-221

The Coconut Industry Board of Jamaica

BASIL O. BEEN
Coconut Industry Board, 18 Waterloo Rd, Kingston 10, Jamaica

1. Introduction

Coconut palms were well established in Jamaica by the end of the 17th century but did not become an important plantation crop until the 19th century. In the late 1920s and early 1930s, Jamaica's coconut industry was chaotic and depressed. Most of the crop was exported as fresh, dry coconuts, and some copra and oil were being produced by independent operators. In response to this situation, the Jamaica Coconut Producers Association was formed as a marketing organisation in 1930. It used members' coconuts to initially produce copra and oil, and later soaps. Legislation was enacted by the Government to protect the local industry from external competition. In 1940, all small producing factories were amalgamated into one company known as Soap and Edible Products Limited (SEPROD). In 1945, the Coconut Industry Board was established as a statutory corporation by Law to administer the coconut industry.

2. Structure, functions and evolution

The Board falls under the portfolio of the Minister for Agriculture. It comprises nine members, four of whom, including the Chairman, are appointed by the Minister for Agriculture, while five are coconut growers elected by coconut growers.

The main objectives of the Board are to organise the orderly disposal of the coconut crop to the best advantage of the growers, and to encourage and assist growers in all aspects of growing the crop. The Board operates on the premise that there is an obligation to find a market for the entire crop, and a duty to encourage the growing of coconuts to meet the needs of the country.

C. Oropeza et al. (eds.), Lethal Yellowing: Research and Practical Aspects, 225–228.

In organising the marketing of the crop, the Board has encouraged growers to establish copra dryers and has provided financial assistance through loans and technical advice in their construction and operation. All copra produced in the island is purchased by the Board as specified by the Law. The cost of transporting coconuts from the farm to copra dryers, and copra from dryers to factory, is subsidised, as are the prices of copra and coconuts for making oil. Thus, these prices are not directly dependent upon world prices. All shares in SEPROD were purchased by the Board in 1945, thereby ensuring a ready market for coconut production.

It became obvious that if growers were to be encouraged and properly assisted in coconut growing, then information based on scientific investigations would be needed. Therefore, in 1959, a Research Department was established and an Extension Division added in 1964 to inform growers of the research findings. Before 1959, research on coconuts conducted by the Ministry of Agriculture was discontinuous. The transfer of the research responsibility to the Board was welcomed as a means of securing continuity and assuring long-term financial support.

Since the Research Department was part of a growers organisation, the research was project-orientated and performed with a sense of responsibility to the growers. As the effectiveness of the arrangement became evident, the Government agreed to give the Board responsibility for plant production, advisory work and all aspects of coconut production. Research performed by the Board covers the following fields: botany, plant breeding, agronomy, crop physiology and plant pathology. Over the years, a vast amount of information has been collected. Technical advice is available from the Board on all aspects of coconut growing.

Aid for specific projects has been readily obtained, both locally and abroad because of the Board's research and development policies. One example was research on lethal yellowing, which was funded by USAID, FAO, ODA, the University of the West Indies and the Ministry of Agriculture. The Board has collaborated with a number of research organisations at home and abroad, and in recognition of the work done on coconuts, many requests are received for advice and consultative services.

3. Services

The Board can be described as a statutory co-operative. It is a non-profit organisation, has no share-holders and any surplus from its operations is used to provide service to the growers. Also available to the Board, is a levy on all coconut products sold locally.

Hurricanes and lethal yellowing have had a profound effect on the local industry and have had a strong influence on assistance programs instigated by the Board over the years. Therefore, the following services are offered.

3.1 Windstorm insurance

After a severe hurricane in 1944, many farmers had insufficient funds to make a fresh start, and replanting was delayed. In 1949, the Board established a Windstorm Insurance Fund under which growers who supply coconuts for copra making, or any other purposes of the industry, are automatically insured against loss of bearing-palms from hurricane. The amount of insurance coverage is based upon the quantity of coconuts supplied. Growers who wish to supplement the automatic insurance to provide a higher coverage per palm, can buy additional (contractual) insurance from the Board. The maximum coverage a grower can have is presently J$125 per bearing-palm by Automatic Insurance alone, or Contractual Insurance alone, or both combined.

3.2 Hurricane Rehabilitation Program

Large-scale destruction of palms by hurricane "Gilbert" in 1988 necessitated vigorous re-planting. The Hurricane Rehabilitation Program was introduced in April 1989. Under the program, seedlings can be purchased for one-sixth of cost and free fertiliser provided for 60% of the seedlings for three years.

3.3 Lethal Yellowing Insurance Scheme

Under the Lethal Yellowing Insurance Scheme, established after lethal yellowing appeared in the main coconut growing region, farmers who supplied coconuts from 'Jamaica Talls' for copra making were given, free of cost, lethal yellowing resistant plants and cash to enable them to replace those palms destroyed by the disease. Once again, the extent of the benefit was related to the quantity of coconuts delivered for copra making.

3.4 Copra Production Acceleration (COPRA) Program

The COPRA Program (1980-1990) superseded the Lethal Yellowing Insurance Scheme and was introduced to make Jamaica self-sufficient in copra in the shortest possible time. Interested farmers were contracted to plant more than 100 seedlings and were required to sell a percentage of any coconuts produced for copra making. Under the agreement, the Board provided the farmer with free planting material, free fertiliser for 5 years, fertiliser at half-cost from years 6 to 7 and advice. Seedlings and fertiliser were delivered to the farm gate.

3.5 Subsidised seedlings

With the presence of lethal yellowing, it is imperative that farmers plant varieties with proven lethal yellowing resistance. The Board is solely responsible for the production of such seedlings which are sold to the farmers at a fraction of the production cost.

3.6 Extension

The Advisory Division of the Board keeps farmers informed of research findings through: farm visits, field training days, resident training courses, booklets and leaflets, agricultural shows, radio and television broadcasts, and newspaper articles. In addition, the Board provides a free foliar analysis service.

Acknowledgements

This paper was published with permission from the Coconut Industry Board, Jamaica.

Coconut product diversification

P.G. PUNCHIHEWA
Asian and Pacific Coconut Community, PO Box 1343, Jakarta 10013, Indonesia

1. Introduction

In Indonesia, there is a saying that the number of uses of the coconut palm is equal to the number of days in the year. An old South Pacific proverb says:

> "He who plants a coconut palm, plants food and drink, vessels and medicine, a habitat for himself and a heritage for his children."

Sayings similar to these are presumably common in tropical America as well, since the versatility of the coconut palm is commonplace (Fig. 1).

The coconut is the premium product of the coconut palm, with the kernel being the key element. It is consumed fresh, or is processed into home-made oil, desiccated, or turned into copra for subsequent crushing to obtain oil.

2. Significance of copra and coconut oil

An estimated 55% of the annual world production of 50 billion coconuts is converted into copra. This remains the principal activity of the coconut industry. Copra was introduced as a source of vegetable oil in northern Europe during the 19th century. New research and the rapid industrialisation of the West led to new uses for coconut oil. For example, in 1832, use of coconut oil improved the quality of candles made from tallow. In 1841, a patent was issued for the manufacture of soap from coconut oil. By 1890, large-scale use of coconut oil for margarine production commenced.

C. Oropeza et al. (eds.), Lethal Yellowing: Research and Practical Aspects, 229–247.

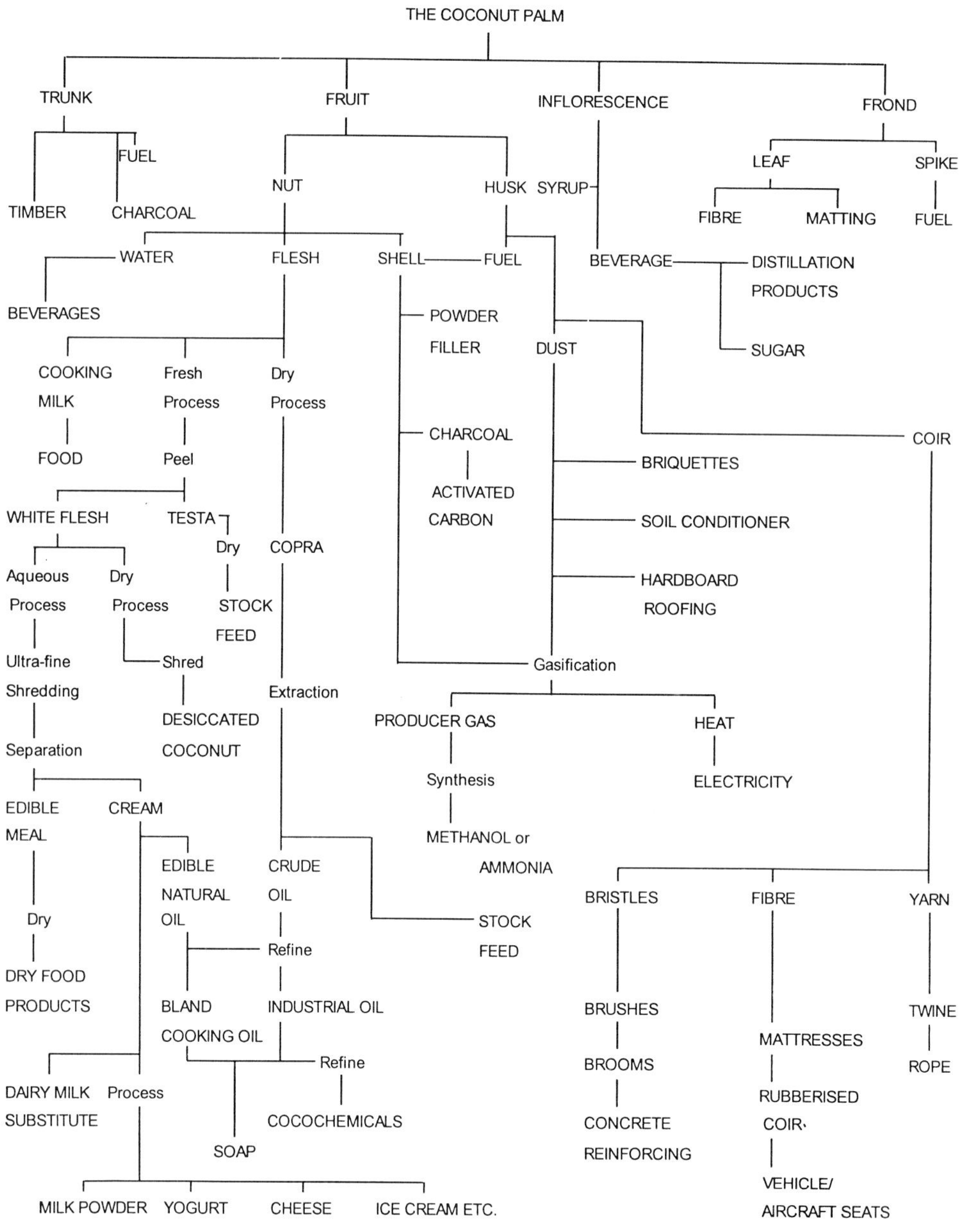

Fig 1. Potential products from the coconut palm [2].

Today, coconut oil is valuable for the production of many food and industrial products such as: cooking oil, shortening, margarine, soap and shampoo. Due to the advantages of coconut oil over other vegetable oils, it commanded a premium price until recently. However, with the increasing availability of substitutable vegetable oils, especially palm-kernel oil, the situation has changed. The prices that copra and coconut oil fetch in the world market have fallen significantly (Fig. 2).

Added to the lower prices, there was increasing domestic consumption and slow growth in production. Therefore, the exportable supplies diminished. Coconut oil has dropped from 33% of the world vegetable oil trade in 1933 to 8% today.

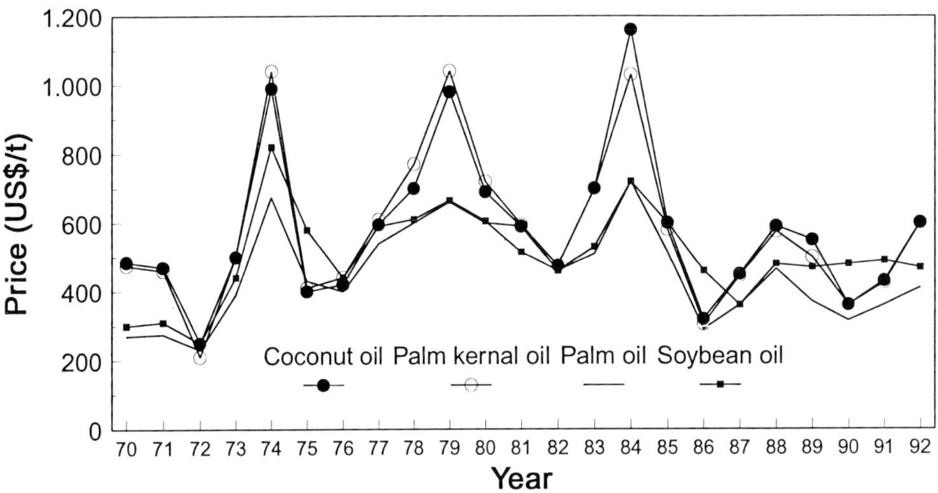

Fig. 2. International prices of vegetable oils form 1970-1992 [1].

3. Need to diversify

It has become necessary for coconut-exporting countries to search for possible alternatives to copra to enhance their foreign revenues. Accordingly, product diversification in the coconut sector is being explored. These countries have recognised the danger of relying on the production of copra alone, although this has occurred at a late stage and with considerable losses.

Vanuatu is one country which relies very heavily on the export of copra. During the period between 1987 and 1991, approximately 45% of their foreign revenue came from copra (Table 1). In spite of production remaining fairly constant during this period, the fall in copra price caused the total value of coconut product exports to drop (Fig. 3). In contrast are India, which only exports fibre products, and the Philippines which exports a variety of coconut products, predominantly coconut oil. A similar drop in value of their respective coconut product exports during the same period was not apparent (Fig. 3). The Philippines has been consistently reducing its exports of copra from an annual average of 1 million tons in 1956/57 to 80 thousand tons in 1991. Instead of exporting copra, they have diversified into a multitude of different products (Table 2).

Table 1. Vanuatu: contribution from the coconut sector to export earnings, 1987-91.

Year	Total exports (million Vatu)	Coconut exports		Contribution from coconuts (%)
		Volume (t)	Value (million VT)	
1987	1502	34125	719	48
1988	1558	30179	953	61
1989	1612	23621	768	48
1990	1606	37297	603	38
1991	1660	32505	549	33

It should be possible for countries to diversify their coconut production by establishing:

a) non-traditional products,
b) by-products,
c) improvement of existing markets, and
d) new markets.

4. Kernel products

As previously mentioned, with 55% of the world production of coconut being processed into copra, there is potential to convert at least part of the copra production into non-traditional coconut food products through the wet process. One such product which is becoming increasingly popular is coconut milk and cream. The technology for coconut milk preservation is well established. Among the producers and exporters of this product are: the Philippines, India, Indonesia,

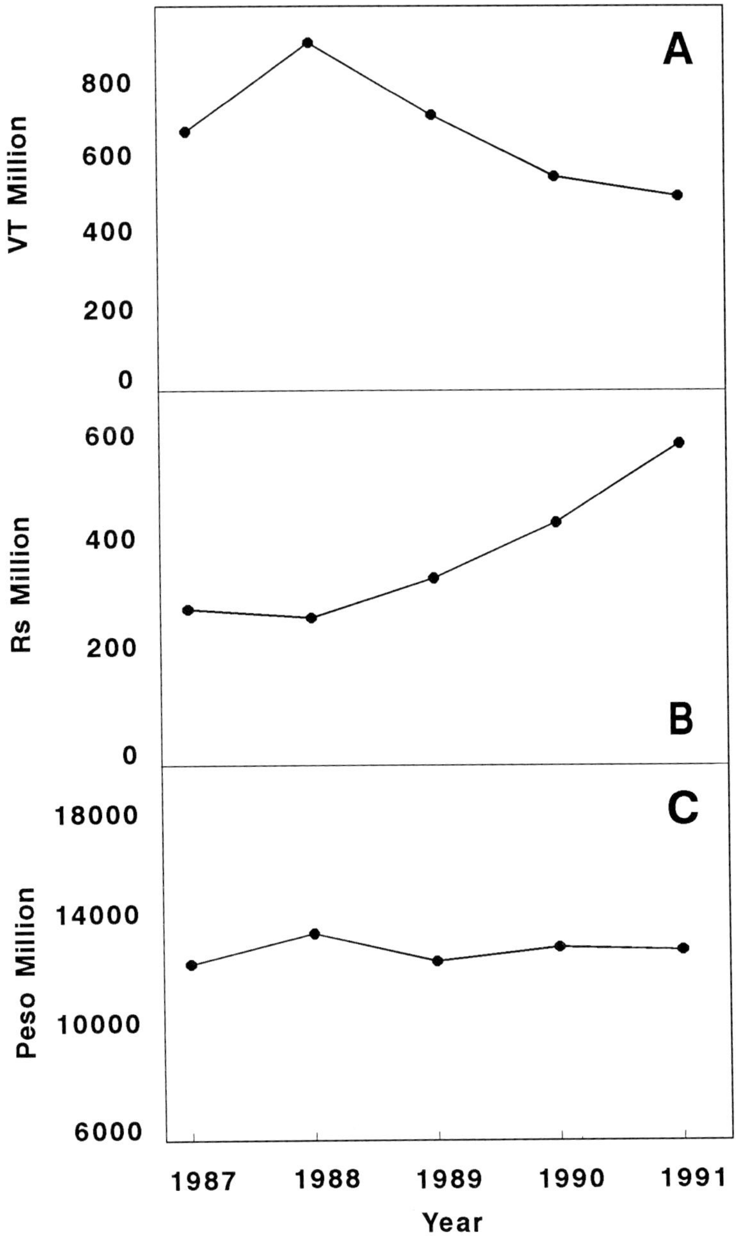

Fig. 3. Total value of coconut products exported, 1987-1991. A: Vanuatu, B: India, C: Philippines [1].

Table 2. Philippines: exports of non-traditional coconut products and by-products, 1987-91.

Commodity	1987		1988		1989		1990		1991	
	Volume (t)	Value	Volume (t)	Value	Volume (t)	Value	Volume (t)	Value	Volume (t)	Value
Chemicals[1]	90706	61676219	85521	72527240	38573	66751040	1015562	49238619	32157	19037254
Crude glycerine	7114	6497993	7285	6163228	5274	5864940	1993	1224797	1233	488813
Refined glycerine	5694	6659874	4900	5600737	4324	5085731	1835	2158746	2678	1644463
Fresh coconut	5629	973119	2851	504961	5803	1582179	7279	1749061	4554	1388697
Acid oil	6523	1327749	8482	2718943	7337	2100949	47 98	1173022	1684	299049
Alkanolamide	1853	1398735	1605	1511096	711	776468	1208	1004229	1931	1448685
Matured coconut	707	144105	484	113234	713	279274	1003	350240	601	281974
Seednut	2081	232595	8	3750	-	-	-	-	-	-
Seedling	1433	361542	498	210498	1441	490025	1127	321809	1054	227236
Charcoal powder	108	3663	32	945	-	-	597	45242	969	135203
Milk	311	413635	221	350450	159	197163	136	181641	531	851395
Hydrogenated CNO	266	241778	100	106240	67	70826	-	-	-	-
Chips	103	137360	98	160829	10	22710	24	43588	335	484878
Cream powder	161	378415	293	713321	43	109207	-	-	506	1598266
Food products	38	28613	391	1984	-	531	-	-	-	-
Fibre	15	2835	-	-	-	-	-	-	-	-
Paring oil	50	30874	41	28815	-	-	-	-	17	10911
Charcoal waste	80	5107	-	-	-	-	-	-	-	-
Toasted coconut	92	64118	-	-	177	256427	70	112654	-	-
Husk	33	11524	47	12190	17	7840	27	10158	147	27338
Soap	-	97	-	-	-	-	-	-	-	-
Activated carbon waste	61	17887	-	-	67	15804	-	-	-	-

Shell	36	2160	10	2220	15	2178	14	1800	86	19600
Water	34	9200	-	-	-	-	-	-	304	254000
Syrup	19	21000	-	-	-	-	-	-	-	-
Toasted rice coconut	3	3015	-	-	-	-	-	-	-	-
Sweetened DCN	11	9135	-	-	-	-	-	-	-	-
Trunk	-	-	82	25494	56	11094	-	-	-	-
Fibre pads	9	4440	-	-	-	-	-	-	-	-
Shell husk	6	1950	-	-	-	-	-	-	-	-
Chopped husk	-	-	-	-	5	1385	12	2400	64	10690
Fibre waste	-	-	37	2828	-	-	-	-	-	-
Coir	-	-	-	-	79	14240	-	-	635	64480
Bukayo	-	-	-	-	3	4557	2	3510	25	33990
Lumber	-	-	-	-	80003	2000	24	2309	-	-
Nata de coco	-	-	-	-	309	336172	200	335301	356	561793
Vinegar	-	-	-	-	243	104656	228	160952	286	200546

[1] methyl ester, fatty acid, fatty alcohol aggregate
[2] copra equivalent
[3] board feet

Malaysia, Sri Lanka and Western Samoa. Western Samoa is a particularly interesting case. It has only 47 000 ha of coconuts and a high domestic consumption and still has three coconut cream processing units. They export an average of 1 300 t of coconut cream per year.

There are many other non-traditional coconut food products that can be derived through the wet process. Nata-de-coco was first produced in the Philippines during the 1940s and has remained a home industry since. Today, nata-de-coco production is becoming more popular because of a high demand on the export market. Another example is coconut milk powder, now being produced in Sri Lanka by a large trans-national company.

5. Coconut sap

There are food products that can be processed from other parts of the palm as well. The inflorescence of the coconut can be tapped to obtain the sap which has a high sugar content. When fermented, the sap becomes an alcoholic beverage which can then be distilled to yield an alcoholic spirit. The sap can also be processed into sugar or vinegar. In fact, Indonesia has made considerable progress in developing coconut sugar manufacturing as a cottage industry. While the majority of this sugar is consumed domestically, approximately 1000 t is exported per month. The export market has compelled the industry to improve its quality and packaging. Another unusual food product which may be considered is the luxury, heart of coconut.

6. Coconut shell

Apart from food products, other parts of the palm can be utilised in the diversification of coconut products. In fact, a former Director of the Coconut Research Institute in Sri Lanka once described the versatility of coconut shell in the following words:

> "In the hands of the beggar, a coconut shell serves as his begging bowl; in the hand of an artist, it turns into a thing of beauty; it provides the humble housewife with a brightly burning fire to cook her simple fare; it lets the chemist unlock its secrets and yield a dozen new things; such is the coconut shell, the most versatile part of the most versatile tree, the coconut palm."

The main use of coconut shell is in charcoal production. Processing of the shell is achieved through a simple process of burning it in a cylindrical drum or pit. This charcoal is subsequently converted into activated carbon, either locally or abroad. Currently, the Philippines, Sri Lanka and Indonesia are the major producers of coconut shell charcoal. In fact, Sri Lanka is now importing coconut shell charcoal from Indonesia and India to meet its requirements for activated carbon. The f.o.b. Philippines price for coconut shell charcoal for the 1987-91 period averaged approximately US$ 190/t. Whereas over the same period, the price of activated carbon from the Philippines was US$ 1050/t f.o.b.. On average, 4 t of coconut charcoal is required to make 1 t of activated carbon.

7. Coconut husk

Another by-product of the coconut palm is coconut husk, and is fully exploited in India and Sri Lanka. In fact, in the Indian state of Kerala, husk processing provides employment for more than 600 000 people. The husk yields a fibre known as coir, which is subsequently used to manufacture rope, mattresses, brushes, cables, upholstery, carpets and rugs. A recent development from the husk is a coir-pith briquette called cocopeat. Cocopeat is made from the residue obtained after fibre extraction. This product is a very good substitute for non-renewable peat and has found wide acceptance in western Europe. Use of cocopeat in those countries thus alleviates the exploitation of peat bogs which have great ecological significance. Formerly, the disposal of the coir residue was also a major problem in coir-producing countries.

8. Coconut wood

Coconut wood is another multi-purpose product. At the Zamboanga Coconut Timber Research Centre in the Philippines, technology has been established for preserving, seasoning and processing coconut timber. This establishment also has training facilities available.

9. Coconut handicrafts

Handicrafts are another non-traditional product derived from coconut palms. Handicrafts can be produced from many parts of the coconut palm, such as: husk, shell, wood, leaf mid-rib and petiole. There is a growing market for these products and the Philippines has been exporting them to the USA.

10. Traditional products: diversification

Diversification of coconut products does not necessarily require the development of new products, but can also involve improving the existing products with minimal cost. Some countries are producing coconut oil in small bottles for body or hair oil. These products are being sold to speciality and tourist markets. Some coconut products can also be used as primary inputs in the manufacture of other products. One example is low technology soap manufacture as a cottage industry. In the Federated States of Micronesia (FSM), their annual production is only 44 million coconuts from 16 500 ha. However, in the early-1980s, they embarked on a product diversification program and have made considerable progress since (Table 3). The experience of FSM demonstrates the potential for diversification of the coconut industry of even very small producing countries.

The possibility also exists to improve the marketing of traditional products, thereby increasing foreign income. For example, the exports of fresh coconuts (mature and immature) is currently increasing (Table 4).

Table 3. Federated States of Micronesia: sales of coconut products, 1989-90.

Product	1989		1990	
	Volume (t)	Value (US$)	Volume (t)	Value (US$)
Laundry soap	37.49	45540	34.23	47711
Toilet soap	7.25	32520	4.01	59329
Shampoo	0.64	3701	1.66	7747
Body oil	6.4	43739	8.4	60881
Suntan oil	1.49	17008	1.15	13397
Cooking oil	1.23	2529	1.11	3556
Liquid soap	0.06	242	0.09	1584
Copra cake	13.01	2580	20.02	3127
Shell charcoal	1.45	1639	3.86	4258
Gift tray-large[1]	1537	9117	2020	13129
Gift tray- small[1]	1892	5343	2462	8870

[1] Volume measured in numbers sold

Table 4. Export of fresh coconuts (number of fruit), 1987-91.

Country	1987	1988	1989	1990	1991
Philippines	3 557 935	2 086 422	5 077 715	8 653 211	not available
Malaysia	43 122 000	53 616 000	78 049 000	89 478 000	56 856 000
Sri Lanka	16 328 371	12 112 027	16 262 630	22 191 224	24 434 297

In exporting fresh coconuts, technology is limited to that involved in packing and storing. Fresh nuts receive higher income than exporting the equivalent numbers as copra or coconut oil. Sri Lankan fresh coconuts are exported to almost 50 countries (Table 5). Therefore, even a non-processed traditional product can generate additional income if new markets are found.

Table 5. Sri Lanka: exports of fresh coconuts and seednuts by destination (number of fruit).

Country	1990	1991	Country	1990	1991
Australia	303 550	88 250	Mexico	20 000	-
Bahrain	310 500	543 360	Maldives	-	390
Belgium	57 500	-	Mozambique	19 375	-
Canada	-	2 000	Netherlands	130 625	18 750
Cyprus	76 000	15 000	Oman	8 980	-
Denmark	-	25 000	Pakistan	6 780 884	6 711 005
France	108 450	2 199	Poland	-	20 000
Germany	298 150	162 000	Qatar	44 747	44 130
Greece	67 500	58 750	Sarawak	130	-
India	23 400	10 000	Saudi Arabia	420 640	520 718
Ireland	37 500	-	Singapore	10 000	10 000
Israel	476 000	393 750	South Africa	1 177 350	1 146 970
Italy	582 875	545 325	Spain	405 200	481 260
Iran	-	10 000	Sweden	93 750	90 725
Japan	10 000	7 200	Switzerland	-	2 500
Jordan	20	-	Syria	103 250	751 900
South Korea	40 100	-	Turkey	55 000	143 500
Kuwait	78 280	580	Caicos Island	15 000	-
Lebanon	563 750	343 250	UAE	5 694 315	7 059 395
Libya	20 000	-	UK	2 695 660	3 310 027
Malaysia	54 000	-	USA	46 250	96 500
Mauritius	1 370 500	1 819 800	**Total**	**22 191 224**	**24 434 297**

11. Small-scale processing

There are substantial possibilities for product diversification by using small-scale technology (Appendices 1, 2 and 3). Sri Lanka is a good example in this respect. Table 6 shows the current profile of the major coconut-based industries in Sri Lanka. These industries are generally located in the coconut-growing areas and are based on traditional technology using locally manufactured machinery and a high labour input. The small-scale coconut-based processing industries are the basis of the rural economy in the principal growing areas. They use about 40% of the total coconut production in the country and process it into a range of products and by-products for domestic consumption and export.

Table 6. Extent of coconut-based industries in Sri Lanka.

Product	Units	Market outlets
Copra	402	domestic and export
Coconut oil	125	domestic and export
Desiccated coconut	58	export
Creamed coconut	1	export
Coconut milk	2	export
Coconut powder	1	domestic and export
Brown fibre	450	domestic and export
White fibre	10	domestic and export
Coir products	125	domestic and export
Shell charcoal	200	domestic and export

12. Constraints

Almost all the processing technologies referred to previously are practised in one or more of the member-countries of the Asian and Pacific Coconut Community (APCC). However, due to differing social and economic conditions, all of the technologies have not been universally applied. For example, in some South Pacific countries, coconuts are not dehusked before the fruit are split and the kernel removed. In such situations, diversification into by-products based on husks and coconut shell is difficult. There are also countries in the Pacific where coconuts are not picked from the palm, but are collected after they fall. Therefore, tapping the inflorescence for sap is not compatible with harvesting operations because tree-climbers are not employed. Labour and collection costs also must be considered. Most of the non-traditional coconut products produced in Asian countries are predominantly exported.

Penetrating non-traditional markets with such products has its own problems and constraints. There are tariff and non-tariff barriers, and duties imposed by importing countries on finished and semi-finished products may be greater than the values of the raw materials. There may also be quality restrictions; a quarantine regulation was recently imposed in Australia on coconut-based handicrafts from the Philippines. Quality control is especially important for food products. For example, a consignment of coconut cream was recently rejected by the USA because it was considered to be below their standards. For that reason, APCC is currently engaged in formulating quality control standards for producing countries. Another difficulty is that information regarding new uses, availability, sources of production and demand are not readily available.

13. Prospects

In keeping with trends for environmentally friendly products, there is a growing demand for non-traditional coconut products. Coconut fibre is a biodegradable product, coconut timber is a renewable resource, activated carbon is required for water filtration, coconut water is a natural drink and cocopeat serves to solve environmental problems in both the producing and consuming countries.

New markets include the niche markets created by expatriates in developed countries who desire the coconut products that were familiar in their home-countries. The tourist industry is another potential market for coconut products as people search for pieces of tropical paradise. Some entrepreneurs are starting to take advantage of this situation. For example, Cocorock of Venezuela are successfully producing and marketing a coconut drink in the tourist destination of Margarita Island.

There is a small amount of coconut product diversification in the Caribbean/ Latin American region that has ready access to the large USA market. Some of these are:

a) grated coconut in syrup, edible coconut oil, coconut pudding and coconut cream in the Dominican Republic,

b) coconut soda and piña colada in Puerto Rico,

c) coconut soap in Mexico,

d) dry whole coconut in Honduras,

e) coconut milk in the Caribbean, and

f) coconut liquor in Jamaica.

It is interesting to note that a small producer such as the Dominican Republic, with an annual production of 170 million coconuts per year, has been able to diversify into so many products. In fact, most of the fresh coconuts imported into Europe today come from the Dominican Republic, Honduras, Sri Lanka and the Ivory Coast. While Sri Lanka and the Ivory Coast were traditional exporters of coconuts, they have been losing market share to the Dominican Republic over recent years. Product diversification is probably why the Dominican Republic has recorded the highest increase in coconut production in North and Central American since 1979 (Table 7).

Mexico is the fifth largest coconut producer in the world, with an average annual production of 1387 million coconuts per year. Mexico's potential for coconut production is vast and the country has the added advantage of close proximity and strong linkages to large markets to the north and south. With these resources, Mexico need not be confined to small-scale processing, but has the potential to develop more sophisticated products such as activated carbon and cocochemicals. With a population of 85 million people, the domestic market for further products should be substantial. However, Mexico still imports some coconut oil, desiccated coconut, coir fibre and even some fresh coconuts.

14. Conclusions

Asian countries have progressed to varying degrees in coconut product diversification. Still, the income generated with these products is low compared to the potential. They also have a long way to go since the high technology and end-uses are still in the developed countries.

Table 7. Coconut production in North and Central America (t).

Country	1979-81	1989	1990	1991
Barbados	2	2	2	2
Belize	3	3	3	3
Costa Rica	25	28	28	28
Cuba	17	24	25	26
Dominica	10	16	12	12
Dominican Republic	69	118	140	150
El Salvador	55	74	78	78
Grenada	7	7	7	7
Guadaloupe	3	3	3	3
Guatamala	3	13	13	13
Haiti	34	36	35	35
Honduras	16	7	7	7
Jamaica	176	200	200	200
Martinique	1	1	1	1
Mexico	851	1157	1091	1101
Nicaragua	1	3	3	3
Panama	23	20	19	20
Puerto Rico	9	7	6	5
St Kitts	2	2	2	2
St Lucia	34	33	33	33
St Vincent	22	21	20	20
Trunidad and Tobago	49	30	40	40
Total	1412	1802	1766	1787

There is a need for coconut-producing countries to diversify. A common attitude is that coconut cultivation is not remunerative, however, it can be. Increased production and productivity, product diversification and proper marketing can change the entire structure of the industry, resulting in increased income to farmers and additional employment in rural areas. This, in turn, may reduce the extent of rural migration to the cities. Diversification will enhance exports and attract additional revenue. It can also eradicate importation of coconut products or their substitutes. Finally, diversification can result in the maintenance of a continuous momentum of activities in the coconut sector.

References

1. APCC (1991) Statistical Yearbook 1991. Jakarta: Asian and Pacific Coconut Community
2. Etherington D and Hagen D (1989) Fresh coconut processing for the south Pacific: fashion and design
3. UNIDO (1990) Coconut Processing Technology Information Document Part 7. Coconut shell products and other products. UF/RAS/78/049. Vienna: United Nations Industrial Development Organisation

Appendix 1. Soap making from coconut oil [3]. Process flow diagram. This is presented for the case of the simplest formula where coconut oil alone (without other fatty oils) is used with colouring and perfuming.

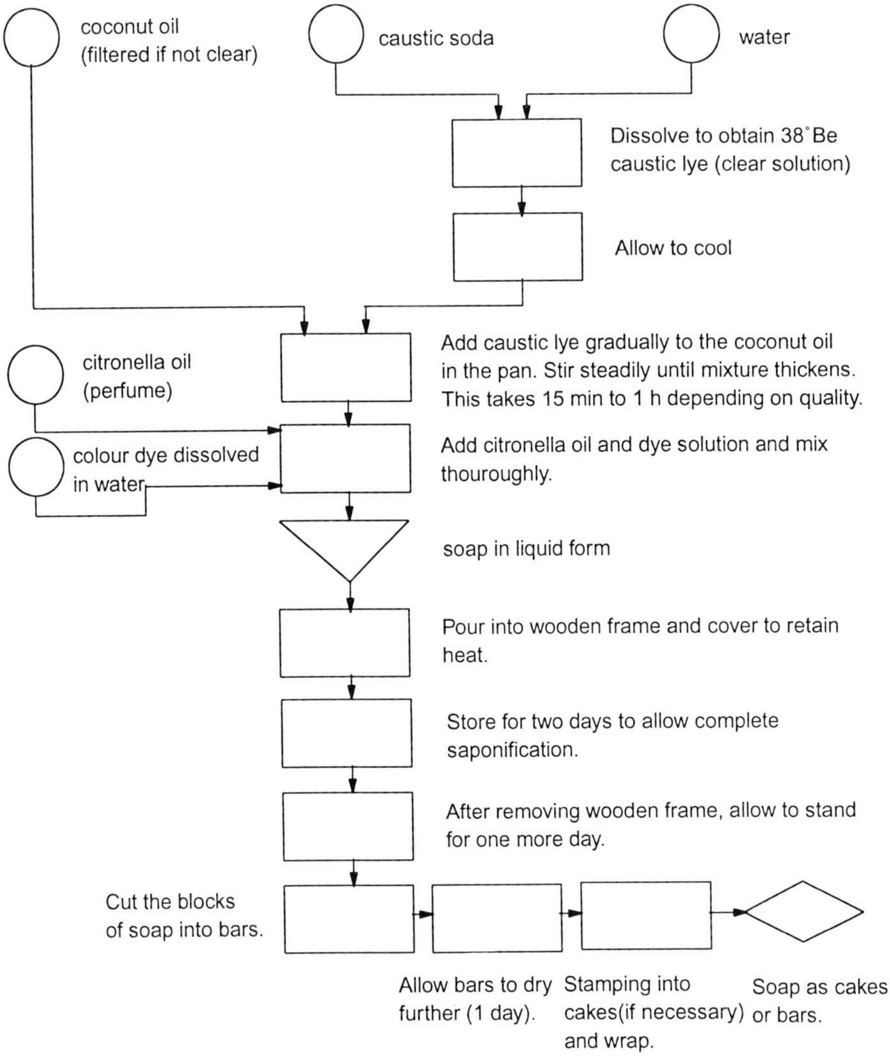

Appendix 2. Production of coconut fibre [1]. Process flow diagrams (simplified).

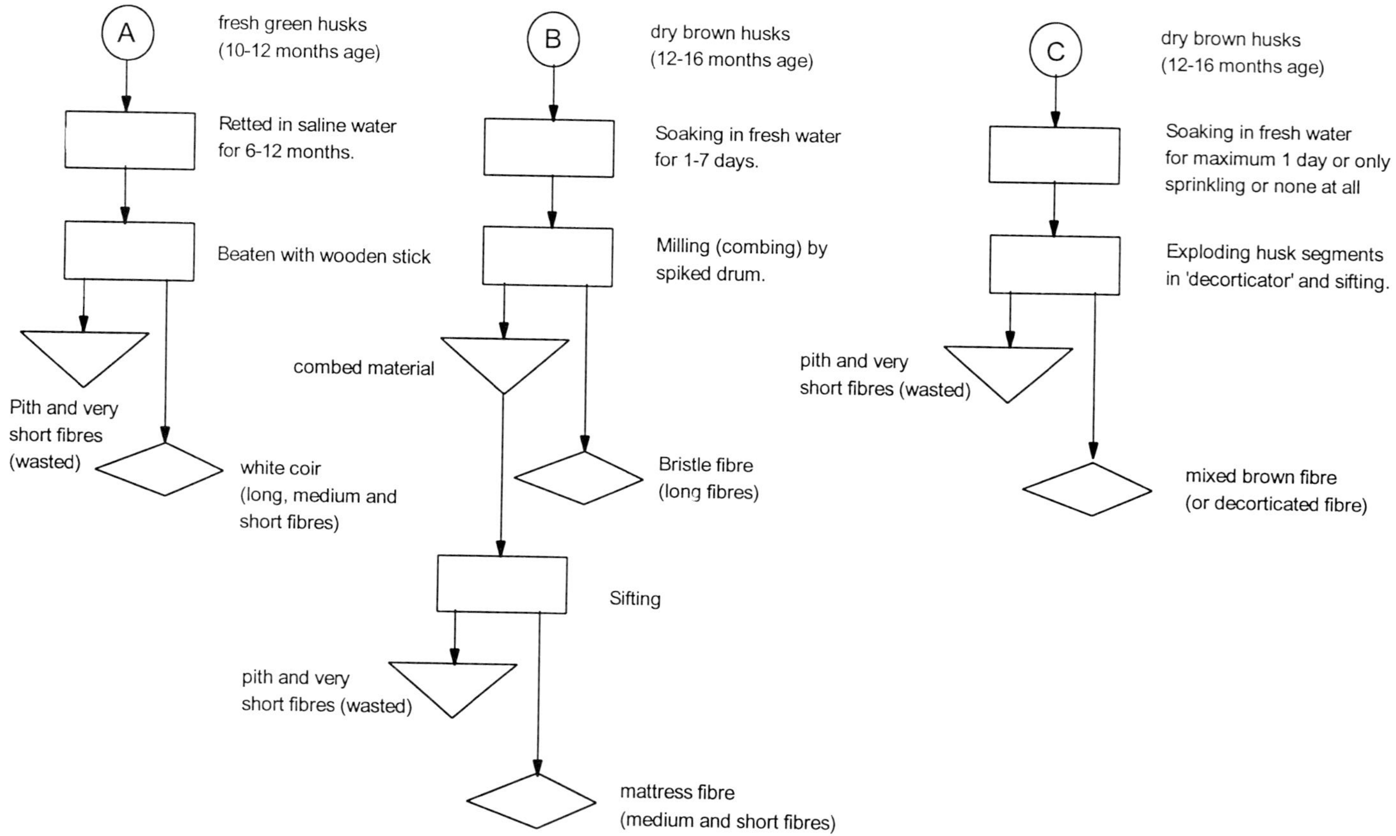

Appendix 3. Manufacture of coconut shell charcoal using simple drum kiln [1].

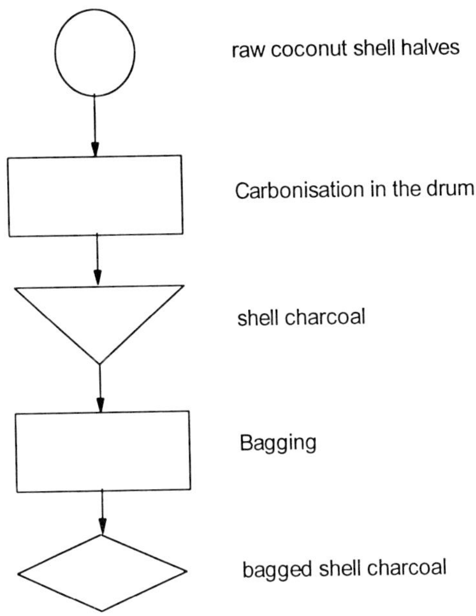

Index

Developments in Plant Pathology

1. R. Johnson and G.J. Jellis (eds.): *Breeding for Disease Resistance.* 1993 ISBN 0-7923-1607-X
2. B. Fritig and M. Legrand (eds.): *Mechanisms of Plant Defense Responses.* 1993 ISBN 0-7923-2154-5
3. C.I. Kado and J.H. Crosa (eds.): *Molecular Mechanisms of Bacterial Virulence.* 1994 ISBN 0-7923-1901-X
4. R. Hammerschmidt and J. Kuć (eds.), *Induced Resistance to Disease in Plants.* 1995 ISBN 0-7923-3215-6
5. C. Oropeza, F.W. Howard, G. R. Ashburner (eds.): *Lethal Yellowing: Research and Practical Aspects.* 1995 ISBN 0-7923-3723-9
6. W. Decraemer: *The Family Trichodoridae: Stubby Root and Virus Vector Nematodes.* 1995 ISBN 0-7923-3773-5

KLUWER ACADEMIC PUBLISHERS – DORDRECHT / BOSTON / LONDON